Water Culture in South Asia
Bangladesh Perspectives

Suzanne Hanchett,
Tofazzel Hossain Monju,
Kazi Rozana Akhter, Shireen Akhter,
and Anwar Islam

Development Resources Press
Pasadena, California

First published 2014
Development Resources Press
P.O. Box 94859, Pasadena CA 91109, USA
www.devresbooks.com

Library of Congress Control Number: 2014913568

Water Culture in South Asia: Bangladesh Perspectives/by Suzanne Hanchett, Tofazzel Hossain Monju, Kazi Rozana Akhter, Shireen Akhter, Anwar Islam
Includes bibliography and index.

1.Water—Social Aspects. 2. Water resources development. 3. Water–Folklore. 4. Environmental health. 5. Arsenic in drinking water. 6. South Asia water traditions. 7. Medical anthropology.

ISBN-978-0-9906337-0-9(hbk)
ISBN-978-0-9906337-1-6(pbk)
ISBN-978-0-9906337-2-3(ebk)
ISBN 978-0-9906337-3-0 (ebk, Smashwords Edition)

Printed by Thomson-Shore
Dexter, MI 48130, USA

Dedicated to
Barbara Hansen and Mohidul Hoque Khan

Contents

Maps

Tables

Figures

Case Studies

Preface

This book idea came out of more than 12 years of team research on Bangladesh water projects. We conducted extensive fieldwork, as a group and as individuals, in the course of conducting baseline studies, midterm and final evaluation studies, and program monitoring for several nonprofit and international organizations throughout Bangladesh and briefly in West Bengal, India. Our principal employers or clients have been UNICEF, CARE Bangladesh, DFID, DGIS, DHV Consultants, Danida, the World Bank, WaterAid Bangladesh, the Government of Bangladesh Arsenic Policy Support Unit, the Noakhali Char Development Sector Project, and the NGO Progoti.

Most of these studies brought us into the homes of rural villagers, who kindly took the time to discuss their concerns and their thoughts about water and its meanings – and their reactions to the serious problem of arsenic in their drinking water. We also drew on our life experiences, friendships with our own neighbors, and our knowledge of South Asian cultural history. As our experience expanded and our focus sharpened, we began to understand that our research offered a unique opportunity to explore the intimate sphere of domestic water management and its broad group of cultural meanings. We felt we had something important to share with more technically-oriented colleagues.

In 2009 UNICEF Bangladesh provided us with funding to do some short-term, systematic research on water culture in two Bangladesh districts, Pabna and Comilla. This support enabled us to organize a series of new interviews that helped us to follow up on key findings and insights. We greatly appreciate the encouragement of our UNICEF program officer at the time, Dr. Richard Johnston, to pursue our research interest. Without that support, this book would not have been possible. Jan Willem Rosenboom, formerly with UNICEF, also has been most helpful with information about the arsenic problem.

An important resource for this research work has been our long-term partnership with the Dhaka survey research firm, Pathways Consulting Services Ltd., Mohidul Hoque Khan, CEO. Pathways not only strengthened our studies with survey research and statistical analysis but also generously offered office space for years on end. Eventually, because of the support of Pathways, our studies combined qualitative and quantitative research methods in ways that our client organizations found quite useful.

Several individuals have contributed to this book in significant ways. Dr. H.K.S. Arefeen has been a constant friend and intellectual sounding board. Dr. Barbara Rose Johnston also has encouraged our research efforts over a period of several years. Pathways staff members Asraul Haque Khan Eitu and Monirul Islam organized all survey work and also trained many junior staff, some of whom went on to collect qualitative information used in this study. Johurul Islam, Ms. Munni, and Ms. Roksana were especially helpful. Our colleague from a 1997-1998 baseline study, Mr. Nurul Absar, retired from the Planning Department, strengthened our early work and collected some interesting material presented in this book. Dr. Md. Faruquee helped us with information on the arsenic problem. Ms. Laila Rahman, with whom we worked on a general review of Bangladesh water and sanitation in 2000, took a number of the photographs that are used here to illustrate our points.

We use ethnographic and other rural studies to supplement our own findings. We especially acknowledge the valuable insights of Dr. Najma Rizvi, Thérèse Blanchet, Dr. Farhana Sultana, Dr. Mahbuba Nasreen, Dr. Sushila Zeitlyn and Farzana Islam, Dr.

Mahmuda Islam, Dr. Jean Ellickson, Dr. Kaosar Afsana, and Dr. Irène Kränzlin, Dr. Clarence Maloney, Dr. K.M.A. Aziz, Dr. Allan Smith, and Dr. Dipankar Chakraborty.

Several people have helped with manuscript review and editing. We especially thank Dr. Stanley Regelson, Dr. David Groenfeldt, Dr. David Rudner, Barbara Hansen, Louise Lacey, Yvonne Champana, Brenda Van Niekerk, and Jon Weinberg for their assistance.

All of our families were helpful and supportive when we needed to be away from home for long periods of rural field-work. We appreciate and thank them from the bottom of our hearts.

Transcription of Bengali Words

Vowels	Bengali Script	Transliteration	Pronunciation & Examples
Short-a	অ	a (or o)*	Awe
Long-a	আ	aa (or ā)*	The a in father
Short-i	ই	i	The i in bit
Long-i	ঈ	ii (or ī)*	The ee in feet
Short-u	উ	u	The u in put
Long-u	ঊ	uu (or ū)*	The u in rude
Short-e	এ	e	The e in bet
Long-e	ঐ	ee (or ē)*	The a in favor
Short-o	ও	o	The o in boat
Long-o	ঔ	oo, ou (or ō)*	The o in photo
Nasalized vowel: ~	ঁ	(Example: ã)	(No English equivalent)
ch	চ	ch	
sh	শ ষ স	sh	
Aspirated consonant		-h	Th, th, Dh, dh, chh, ph, bh
Retroflex-t/th	ট ঠ	T	*brishTi* ('rain')
Retroflex-d/dh	ড ঢ	D	Dhaka
Retroflex-n	ণ	N (or ṇ)*	

Retroflex-r	ড়	R (or ṛ)*	*baaRi* ("homestead")
ng	ং	ŋ, ng (or ṅ)*	*gaŋgaa* (Ganges River, or the Hindu goddess)

*Alternate transliterations are used in some citations.

1. Introduction

This book concerns the cultural frameworks that surround water development projects. Such frameworks are embodied in practices whose importance is not always recognized by the scientists in charge of the projects. But they do influence outcomes. We view people in developing countries as having their own points of view and their own ways of life—as people who recognize many aspects of their environment and the ways it sustains human communities.

Water is the same chemical compound, H_2O, anywhere it is found, but ideas about water differ from one world region to another. Understanding cultural beliefs and feelings can help scientists, engineers, health specialists, and other development professionals to create sustainable change.

Such understanding does not come easily, however. Epistemological and other differences pose daunting challenges to communication. When a scientist or an engineer encounters indigenous or folk views and devises easy explanations for them, confusion and even conflict may result. Professional training rarely, if ever, prepares him or her to integrate non-scientific approaches into development projects.

Currently accepted scientific theory and knowledge have become established mostly during the past 100 to 250 years. Chemistry and bacteriology, for example, have undergone continual revolutions during this period. In the 1770s, Lavoisier developed a radically new system of chemistry by establishing a simple definition of a chemical element. The germ theory of disease was verified by Louis Pasteur and others only in the 19th century. The people that development professionals try to help, on the other hand, may be using ancient ideas such as humoral theories of medicine, purity and pollution, or totemism.

A fascination with water is universal among human communities. Not only is it essential to life, but water also has properties that inspire metaphorical and poetic thought. Water's absorptive properties inspire hope that it can remove spiritual problems or sin as well as ordinary dirt. Water's flow is compared to human growth and the movement from one generation to another. Water bodies and rivers can define environments and identities. One scholar, Veronica Strang, finds that "the meanings poured into water have proved highly consistent over time..." She refers to humanity's "highly complex relationship with water," a relationship "in which physical, sensory and cognitive experiences articulate with cultural meanings and values." (Strang 2004:3)

Australian Aborigines have well-documented mythological traditions connecting them to water bodies, animals, or other features of the natural environment, which in some groups are claimed as totemic ancestors. Monica Morgan, a member of an Aboriginal group in the Murray Darling River Basin of southeastern Australia, describes her people's feeling about their place and its water, as, "We have always been, and will always be, the First People of this land. We belong to it, and the water that flows through our country is as the blood that flows through our veins. Our bodies are formed from the country and remain tied to its rhythms..." (Morgan 2012:454-55)

The geographer Jamie Linton argues persuasively for the need to see water as more than a commodity. Some scientists and business people view water, he says, "…as just another 'resource'—something to be captured or tamed, put in containers or otherwise diverted from its natural path, and transported far away to be used and sold for money…. To First Nations People, however, water is seen very differently. A creek, which to a non-native person may be seen simply in terms of flow rates and acre-feet per year, may have a special name and spiritual significance. It may be a private bathing place for special ceremonies or initiation rites, or in some cases be owned by a particular individual or family. It not only physically and spiritually cleanses people, but it also cleanses the earth and eventually, the sea to which it inevitably flows, if left alone." (Linton 2006, citing a 1992 talk by Chief Cathy Francis of Canada)

In Bangladesh villagers have their own cultural approaches to almost all aspects of life, water included. Many Bengali-speaking people in Bangladesh and India, for example, believe that their civilization is based on water bodies and rivers, and that people who live next to rivers and water bodies have unique opportunities to be prosperous. Water language, concepts and symbols have a central place in this way of life. According to some of the people we have met, "Another name for water is life." Culturally-based water knowledge and practice have helped people to understand and cope with their environments over the centuries. The old ideas have been tested are trusted for their practical utility. They also reflect and connect with the social and moral principles at the heart of social life. New ideas often are welcomed and generally discussed as possible ways to improve local health and well-being, but the well-established, older views still are trusted to a large extent.

The right to water now has been enshrined in two United Nations declarations. The first declaration was made in 2002, and in 2010 the rights statement was expanded to include "safe and clean drinking water and sanitation," deemed essential to a satisfactory quality of life.* As important as these declarations

*The right to water was formalized in General Comment No. 15, by the UN Committee on Economic, Social and Cultural Rights. This Comment provided a rights emphasis for the 2005-2015 UN International Decade for Action on "Water for Life." The second step was made in July 2010, when the UN General Assembly adopted a resolution that "recognized the right to safe and clean drinking water and sanitation as a human right that is essential for the full enjoyment of life and all human rights." (A/RES/64/292 of 28 July 2010). Shortly thereafter, in September 2010, the UN Human Rights Council further confirmed that it was legally binding on states to respect, protect, and fulfill the right (A/HRC/15/L.14

are, however, they do not ensure that all people will have such access.*

Water is a prominent topic of discussion in the United Nations "post-2015" deliberation process, the goal of which is to define international standards to guide future development efforts. A new set of "sustainable development goals" (SDG) is likely to replace the currently used Millenium Development Goals (MDG) in 2015. Several concerned organizations, including the United Nations Water organization (UN-Water), are arguing that the SDG's should include one new goal dedicated entirely to "securing sustainable water for all." (UN Water 2014, Leone 2014)**

International and regional water development activities are almost all organized by scientifically oriented professionals. However, there are fundamental differences between scientific thinking and the indigenous or folk views typical among the people whose water is being "developed." Physical and biological science is based on the assumption that the empirical method and hypothesis testing—guided by evolving theory—is the best way to build valid knowledge about any subject. Scientific thinking needs to be narrowly focused on well-defined data units. Indigenous or folk thinking, on the other hand, tends to be more holistic. While indigenous knowledge specialists have

---

of 24 September 2010).

*While these rights declarations do not entirely solve the world's domestic water supply problems, "These major policy shifts have been heralded...as a move ... toward addressing global water inequalities." (Sultana and Loftus 2012:1) These declarations, however, are so broad that they can be used to justify privatization and commodification of water resources by moneyed interests that deprive ordinary people of vital water resources. So there is a quest underway for new frameworks or paradigms that can supplement the UN declarations and preserve access to water needed to sustain human life.

**. This deliberation process is going on at the time of this book's publication. In July 2014 a stand-alone Sustainable Development Goal (number 6 out of 17) was proposed to the United Nations General Assembly to "Ensure availability and sustainable management of water and sanitation for all."

experimented and produced useful inventions (herbal remedies, for example) over the millennia, their knowledge usually is constructed in ways that differ from the scientific type. For example, indigenous knowledge typically makes connections between physical and moral phenomena, between the spiritual and human worlds, and/or between social action and individual physiology.*

Doing fieldwork on behalf of development agencies since the early 1990s, we often are in the position of telling the professionals how their efforts affect local life and how effectively people understand (or misunderstand) project messages. We have lived inside the science *vs.* folk knowledge gap for more than ten years. We find development professionals mostly sympathetic and well-meaning, but they may balk when pressed to understand the culturally-based views of the people they are trying to help. Processes of communication need to change in ways that respect the identities and histories of intended program beneficiaries. We will suggest ways that scientists and engineers can make use of a cultural context to create sustainable development projects, whether in Bangladesh or elsewhere.

Among scientists and engineers, there is a common tendency to regard folk beliefs as based in religion, as signs of poor education, or simply as superstitions needing correction. The scientist's powerful position in the typical development project supports his or her sense of entitlement to this view. An understanding of the culture concept, however, will increase chances of development success.

Culture: A Working Definition

"Culture" is a frequently mentioned but often misunderstood concept. Every human being (including the scientist) that has grown

---

*A detailed discussion of the differences between indigenous and scientific knowledge can be found in the book, *Indigenous Peoples and the Collaborative Stewardship of Nature*, by Anne Ross *et al.* (2011).

up with social connections and learned a language has a cultural orientation of some sort.

We use this definition: Culture is the body of principles, rules and values that guide human choices. These are not always conscious. They are basic to a person's identity and to social and emotional life. They endow human actions with perceived meaning. They validate social arrangements. Culture "is learned and acquired by individuals, but has a transgenerational quality beyond the lifetime of individuals." (Rosman and Rubel 1981:6) Cultural principles explain how the world works, which responsibilities go with which social roles, what the signs of health or illness are, and generally why certain ways of doing things are better than others. There is a sense that one's own cultural principles are normal, natural and right. Other people's cultures and languages, however, tend to feel uncomfortable and unnatural.

People with a common cultural framework speak the same language, both literally and figuratively. With language come semantics, gestures, stories, proverbs and other meaningful verbal, non-verbal and artistic communication or expression techniques that integrate, explain, or justify connections among the spiritual, moral and physical aspects of life.

Symbolic actions and discourse—all based on cultural principles—are at the heart of day-to-day social life and the definition or redefinition of core values. Language, discourse, and symbols—these cultural elements are human beings' tools for continually producing, reproducing and negotiating every aspect of community life. They also provide a framework for environmental adaptation.

There is considerable variety in the ways that people interpret and act upon their common cultural heritage. Culture is a set of *principles* or *assumptions,* a world view or way of thinking, not merely a set of customs or practices. People are creative in their use of cultural parameters, so there is always variation in actual practice among those who share a cultural framework. People can and do disagree on how their common cultural principles apply in specific situations, but the principles themselves are taken for granted. While common and strong, culture is not rigid. Culture can and does change, though slowly. Furthermore, people can and do use more than one cultural framework, as when they travel back and forth between different countries or ethnic communities.

Studying cultural phenomena is done in various ways, but it always involves considerable observation, conversation and empathy. The goal is to understand people's approach to life and their assumptions about what various acts or events mean. Interpretation depends in part on observing behavior and thinking patterns. Symbolism and language are crucial to cultural analysis.

The Relevance of Culture to Water Resource Management and Domestic Supply Programs

We agree with Brugnach and Ingram, who argue that, "Failing to address the biological and cultural diversity associated with water problems is no longer suitable. Instead, a cross-cultural approach that encompasses diversity is needed." (Brugnach and Ingram 2012:61)

There are different kinds of development activities involving water. One type is concerned with the management of water resources in large regions such as watersheds, river basins, wetlands, and lands over aquifers. Another type relates to domestic supply—that is, ensuring access to safe water for cooking, personal hygiene, and other home-centered purposes. A special focus within domestic supply is known as WASH, or Water, Sanitation and Hygiene. Cultural issues figure differently in these different types of water-related development activities.

Cultural Diversity and Large-Scale Water Projects

Water is urgently important to environmental sustainability, and culturally diverse populations often depend on the same sources. If the flow of a river is blocked to irrigate the fields or dammed to support hydro-electric power needs of one region, for example, populations downstream are deprived of water. Human needs compete and often conflict. Ecological balance is at stake. Cultural issues have received considerable attention in these kinds of projects. Hundreds of large-scale water projects by now have confronted the need for culturally diverse groups of stakeholders to negotiate and compromise. Indigenous peoples in Australia, North America, South America, and elsewhere have struggled mightily to protect treaty rights and their access to water in their home territories against infringement or usurpation by more dominant groups.

Procedures exist to negotiate among economic, social/cultural, and environmental needs in such large-scale water resource management situations. Prominent among them are Integrated Water Resource Management (IWRM)* and Environmental Flow Analysis.** These processes are complex and political, not mere

*IWRM is "a process which promotes the coordinated development and management of water, land and related resources in order to maximise the resultant economic and social welfare in an equitable manner without compromising the sustainability of vital ecosystems." (Global Water Partnership 2000, quoted in Matthews 2012:358)

**Environmental Flow Analysis (EFA) includes more than 200 different assessment methods. The general purpose is to determine the minimum flow required to sustain a healthy river system. "The role of EFA is to relate these hydrologic characteristics to physical and ecological responses, and thereby inform restoration of the socially valued benefits of biodiverse and resilient fresh-water ecosystems through participatory decision-making informed by sound science." (Arthington 2012 & Poff et al. 2010) "Environmental flows describe the quantity, timing, and quality of water flows needed to sustain freshwater and estuarine ecosystems and the human livelihoods and well-being that depend on these ecosystems." (Brisbane Declaration 2007, quoted by Arthington 2012)

technical exercises. They are widely used and respected by now, and they increasingly include consideration of needs of diverse cultural groups. The Ramsar Convention to protect wetlands is another example of an integrated approach to large-scale water resources management. In 2002, the Standing Committee of the Ramsar Convention adopted a resolution regarding its designation of Wetlands of International Importance: that "incorporating cultural values in the management of wetlands" would be its policy henceforth. (Papayannis and Pritchard 2010)

Although these procedures are widely accepted by now, the literature on water resources management includes many complaints about discrimination and inequality in negotiations. One quote sums up the epistemological and power problem from the marginalized group's point of view: "Indigenous knowledge, where it differs from science, is regarded as inferior." (Ross *et al.* 2011:273)

Despite the persistence of such problems, there are by now a number of cases in which indigenous people have managed to establish themselves as co-managers of water and other resources in their territories with some degree of dignity and respect. An agreement in the Murray Darling river basin of southeastern Australia is one example. A coalition of indigenous groups negotiated a Memorandum of Understanding in 2007 with a governmental commission that specifies that, "...Cultural... and social values should be given equal status with economic values when policy and management decisions are made." (Darling 2012:457) Native American tribes and First Nations also have forged successful co-management agreements that enable them [to] exercise significant autonomy over their lands and waters.*

*A recent example of this trend is Haida Gwaii in British Columbia, Canada, where cultural rights, livelihoods, and stewardship have been strongly linked to sustainable watershed and "marinescape" management as well as biodiversity conservation through new environmental governance arrangements. (Thornton 2012:135)

The formal purpose of such negotiations is to set up planning processes and devise policies that level the playing field. If it works well, the process will represent and respect interests of marginalized and indigenous groups. Respect for cultural diversity is all-important. Power differences always are involved. (Hiwasaki 2012:527) Details of culturally based belief and practice are less important than respecting the rights of marginalized groups.

Cultural Factors in Domestic Supply and WASH

Details of cultural heritage figure differently in projects that are concerned with domestic water supply and WASH. Access of diverse cultural and social groups to safe water is an important issue, as it is in large-scale works. But cultural views of water need more attention in this domain, because home-based ways of using water have serious health implications. Sanitation practices, especially management of feces, affect water safety. Domestic supply activities, therefore, need to be linked to WASH services.

People manage their personal habits, customs and community practices according to their own cultural principles, at least to the extent that circumstances allow. The most relevant cultural materials are language and popular ideas about health and illness. Water's importance in daily routines needs to be explored. Water's customary uses in healing and life-cycle ceremonies also deserve attention. Myth and folklore also may be relevant.

The importance of culture is frequently acknowledged in WASH programs. As one online document points out, "… Respect for, assessment and integration of the cultural context of users such as religious or cultural beliefs, gender or generational differences in water and sanitation programmes
are crucial to mitigate failure risks and promote sustainable solutions." (Tratschin n.d.)

This insight has come at a high cost. Failed experiences have forced scientists, engineers, and public health specialists working on domestic supply and WASH projects to rethink their efforts. One project in southern Africa, for example, found people rejecting water from sand filters, even though such filters seemed completely logical and could help to reduce water-borne diseases. They were cheap, did not use electricity, and could be made from locally available materials. The development agents, however, had failed to recognize the locally perceived importance of drinking only *running* water, not standing water. "They were used to only drink running water because in their experience water in stagnant pools was not doing any good to their health. And because water coming from a sand filter stands for a day while being processed, the users refused to drink it…." (Tratschin n.d.) Similar problems plague the sanitation field. Latrines—though essential to prevent contamination of water sources—may or may not be used to contain human fecal waste, depending on local culture and socially meaningful practices associated with elimination. They may be used for laundry, storage or other household purposes instead.

Bangladesh has had a leadership role in integrating cultural principles into WASH programs. In the early 1990s, for example, CARE Bangladesh and the International Centre for Diarrhoeal Disease, Bangladesh (ICDDR,B) started a ten-year model project eventually called SAFER (Sanitation and Family Education Resource). This project was unusual in focusing entirely on personal communication, rather than on "hardware," meaning physical equipment. Through a network of NGOs (non-governmental or non-profit organizations) coordinated by CARE, program participants were engaged in problem identification, design, monitoring and program evaluation. It was an intensive effort that included weekly sessions with NGO staff and frequent checking by CARE managers. The project produced a large number of educational materials—flash cards, games, posters, and so on—that

Photo 1-1. A CARE-SAFER self-monitoring group promoting latrine use. Chittagong District, 2001 (Photo credit: Suzanne Hanchett)

were tailored to the tastes of diverse ethnic and socio-economic groups. These materials were widely disseminated and further adapted by virtually all other WASH practitioners in the country. CARE-SAFER was internationally recognized as a trail-blazer. (Bateman 1995, CARE Bangladesh 2001)

Adjusting Scientific Information to Local Needs

Development projects rarely try to change everything about a place, but they do need to change certain things. They focus on a specific goal, such as reducing the spread of water-borne diseases. Specific objectives arise from this goal: for example, promoting consumption of safe drinking water, hand washing at certain times, or use of sanitary latrines. Folk culture need not change entirely to achieve these objectives, but some aspects definitely need attention. Success depends to a large extent on how well accepted the new, scientific information is—whether people are willing to make the personal and social changes needed to integrate a new idea, practice, or technology into their daily routines. Not knowing details of local culture and/or not respecting it can lead to surprising results, including failure.

The purpose of recognizing folk or indigenous views in water projects (of any type) is not to glorify old traditions or bring back some imagined environmentally ideal past. It is to establish the right of people with diverse histories to have a say over how their resources are managed, and to accept the fact that diverse points of view—different cultures—do exist. Power differences between development agents and the people they are supposed to help will interfere with communication unless this principle is sincerely accepted. The call is *to accept and respect cultural diversity*. Accepting this principle will require re-thinking the way that development projects are planned and implemented, but it will have long-term benefits.A critical issue is the failure to include women in much water-related planning. Most water projects, especially large-scale projects, are implemented by government agencies in Bangladesh and other developing countries. Government officers are mostly educated men.

Respecting housewives' views—or even meeting women and speaking with them as equals—will require a significant change in the typical official's approach. Whatever the obstacles, scientifically trained development agents *must* communicate with women, especially about domestic water supply issues, at all stages of project planning and implementation, if they expect their development work to be effective.

The Bengali Study Population: History and Environments

Our research focuses on Bengali-speaking communities. Bengali-speaking populations nowadays are concentrated in two countries, India and Bangladesh. Their common origin is the old state of Bengal, an area with one dominant language, Bengali. This state was partitioned in 1947 at the end of the British colonial period. The western part became the Indian state of West Bengal, and the eastern part of Bengal became East Pakistan. In 1971, East Pakistan separated from West Pakistan (now Pakistan) after a bloody civil war and became the independent nation of Bangladesh. (Map 1-1)

Political turmoil has led to large population movements during the past century. The first big change was the 1947 partition of India and Pakistan. The second was the 1971 civil war that resulted in the formation of Bangladesh as a new nation. These changes strongly affected some of the areas covered by this study, especially the Comilla District, which is near to the eastern border with India's Tripura State. At partition, Hindus left for India, and Muslims arrived. Some exchanged their properties with others in the new country, and some just fled. This process has, of course, slowed down, but it has not completely ended. The civil war had devastating effects on all areas of the country, as did a series of political upheavals that followed it. While a village or small town may seem long-settled, many of the families are likely to have come from elsewhere within the past generation or two. Social tensions, pressure on resources (including water), and disrupted or new social networks are typical. According to the *Index Mundi*, the population of Bangladesh in 2004 was 89.5% Muslim, 9.6% Hindu, and 0.9% other religions.

There are eight Bangladesh districts covered by most of this study. (Map 1-2) They represent a distinct set of environmental

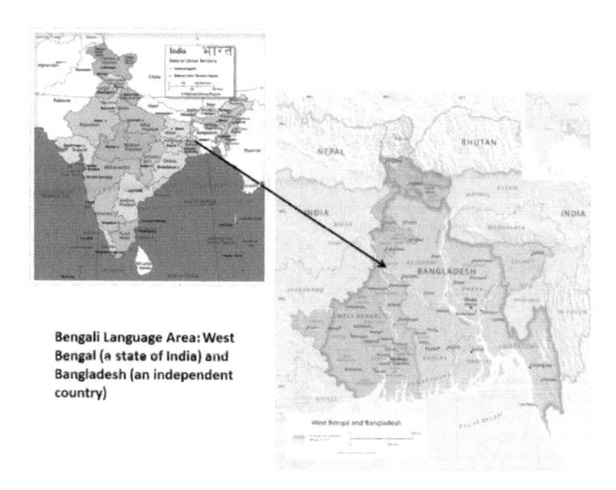

Map 1-1. Bengali Language Area of South Asia (Credits: India Map, Nations Online Project; West Bengal and Bangladesh Map, Anandaroop Roy Cartography)

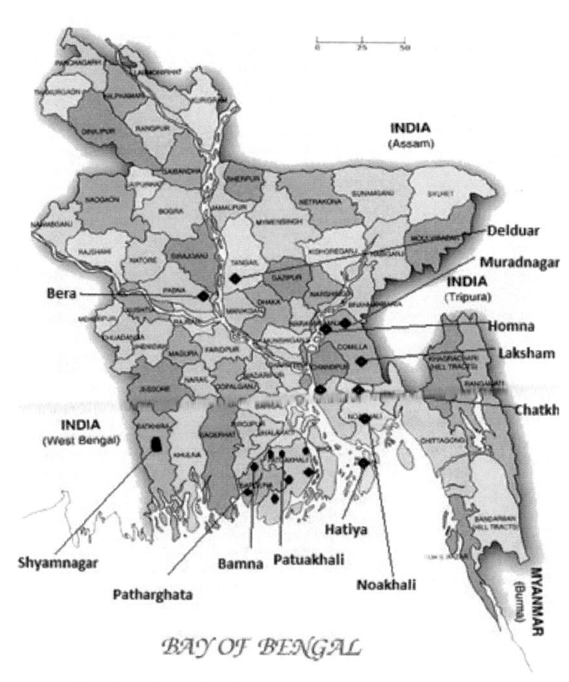

Map 1-2. Principal Bangladesh Study Locations (Details in Appendix 1)

Photo 1-2. Village women sat on the floor at a water resources project meeting in Tangail District, ca. 2004.

Photo 1-3. Village men and government representatives at the same meeting sat on chairs at a table. (Photo credits: Suzanne Hanchett)

conditions. Eastern and southeastern districts (Tangail, Comilla, Laksmipur, and Noakhali Districts) are in relatively low floodplain areas, prone to annual flooding three or more feet deep. Other places (Pabna's Bera Subdistrict and parts of Noakhali, Laksmipur, and Comilla Districts) are even more deeply flooded in the rainy season. None of our locations is far from a large river or canal system. The southern districts of Patuakhali and Barguna and the mangrove forests of Satkhira District are located in or near the Bay of Bengal and strongly affected by tidal flows and cyclones. Villages and towns alike in these southern districts are regularly inundated with tidal waters. Aquifers providing tube well (bore hole well) water for drinking and cooking tend to be very deep (below 350 feet and often as deep as 900 feet) in the coastal belt, so tube wells are more expensive to install and more distantly spaced than in other districts

These southernmost subdistricts face a problem of salinity in their surface and underground water sources. Large and small ponds—sometimes called tanks—are found in all of our locations, and the people make much use of pond water for domestic purposes. The southern environments are fragile and vulnerable to climate change. Some other parts of the country, where we have not conducted research, have different types of environments. They may possibly have some different cultural approaches to water.

Bangladesh's large network of rivers, streams and canals totals at least 15,000 miles (24,140 km.) in length. "They consist of tiny mountain streams, winding seasonal creeks, muddy [canals], some truly magnificent rivers and their tributaries and distributaries," as the geographer Haroun Er Rashid explains. All of them, except those of the far southeastern Chittagong region, belong to one of three major river systems, the Ganges (Padma), the Brahmaputra/Jamuna, and the Meghna-Surma. (Rashid 1991:44) In terms of breadth and total annual volume, the Padma/Ganges-Lower Meghna is the third largest river in the world, smaller only than the Amazon and Congo systems. (Rashid 1978:56) The lives of people we have met are strongly influenced by these huge river systems.*

Water Scenarios

The Bangladesh Delta's three major rivers enter the country from India and flow into the Bay of Bengal. Wetlands (*bils*) and oxbow lakes (*baors*) are abundant and support large populations of fish and other wildlife. Although agriculture still is the country's primary occupation, ample surface water supports fishing and fish culture industries. Much of the plains area is less than ten meters above sea level.

Distribution of water in this landscape changes dramatically according to the season. Monsoon rains produce floods, which are mostly benign and good for fisheries and agriculture, but destructive floods, tornadoes, and cyclones also are part of the picture. Cyclone shelters are now found throughout the southern coastal belt districts. Water flow decisions are made in India (to dam or release river water) in ways that create drought or floods in parts of Bangladesh. But Bangladesh has not yet been invited to participate in the meetings where these decisions are made.

Much of the land in Bangladesh is unstable. There is a large population living on sand-bar islands (*chars*) and fragile river banks. These lands are continually destroyed and re-shaped by erosion, especially during the rainy season. Silt accretion forms new lands that extend southward out into the Bay of Bengal, and people rush to settle them, fight over them, and so on.

Over the years, the complex water-related processes of this delta region have inspired huge engineering projects, most of

*Rashid estimates that in an average year 870 million acre-feet of water flows into Bangladesh from India, and there is an additional 203 million acre-feet of rainfall. After deducting evaporation, evapotranspiration and deep percolation, some 953 million acre-feet flows out to sea. One million acre-feet equals 1233.6448 cubic meters. (Rashid 1991:43-44)

which are intended to protect agricultural production and urban infrastructure. Engineers' efforts, however, sometimes have had unexpectedly negative consequences for fisheries and even agriculture. Man-made flood protection structures, such as river embankments or raised structures (polders) enclosing agricultural fields, are found in all regions. These structures can interfere with drainage and tend to create serious waterlogging problems in the southwest regions. Water control structures are not always used for the intended purposes. People may settle on embankments, for example. Farmers may cut embankments illegally, in order to let water in or out. Stimulating local interest in maintenance of government-constructed infrastructure is a major concern among development agents involved in water resources management.

We have observed four major trends in the overall water scenarios of the places covered by our field studies. First, there has been a steep decline of open water fisheries and availability of wild fish. Natural fisheries have declined as a result of overfishing and blockage of water inlets by embankments, poorly planned roads, and newly constructed settlements. Use of chemical pesticides in rice cultivation also has contributed to this trend.

A second trend is increased leasing-out of open water bodies that were once regarded as commons, or resources available for anyone to use. All or most have been freely available for use by professional or traditional fisherfolk and the general public until recently. In recent years, there has been more and more leasing-out of marshland (*bil*) fisheries, and even of fishing rights in sections of rivers and canals. This practice is a source of revenue for local government. The end result is that fishing rights to many water bodies now are privately controlled.* Fishing is

*It is often mentioned in lease documents that local fishers will be allowed access to leased-out water bodies (wetlands, ox-bow lakes, and others), usually referred to as *jalmahaal*. In practice, however, local elites fully control these leased areas with support from government officials.

an important source of food for poor families, so this change has produced considerable hardship.

One man in Pabna District, a retired union council secretary,* explained the situation in his region: "The Atrai River is a valuable source of bathing and cooking water here, and also fishing," he told us. "This section of the canal extending out from the river has a large number of fish, and it remains full of water year-round. The whole canal has been leased out by the government for fishing purposes to 22 people, and some parts of the Atrai River also have been leased to rich people for fishing. The general public can use the canal and river for bathing and collecting cooking water, but fishing is not allowed. It is the common story here that most water bodies, especially large ponds, the river, and the canal, are leased out to rich and influential people. ...Only officially registered people (with lease rights) can catch fish.... The elite group... is controlling the resource and selling fish to members of the traditional Fisherman caste!"

This watery version of enclosure of the commons poses a serious threat to poor people's access to both water and fish. People who own or lease water bodies used for fish culture are restricting others' access to these formerly open sources in new ways in many parts of Bangladesh. This problem is not limited to Bangladesh, of course. As several authors have discussed by now, ownership of water bodies and privatization of water supplies are reducing access to formerly available sources in many countries.

The commons view of water is described by Bakker (2012:30) as an approach that "asserts its unique qualities: water is a flow resource essential for life and ecosystem health; non-substitutable and tightly bound to communities and ecosystems through the hydrological cycle."** Technology and science can support the

*A union council (union *parishad*) is an elected body representing a population of 20,000-30,000. The union is the lowest level of Bangladesh government.

**Bakker uses a cultural argument in support of the commons view: "Water," she argues, "has important cultural and spiritual dimensions that are closely articu-

privatization trend, or they can adopt a more balanced, long-term view.

A third trend, associated with the decline in natural fisheries, is an increase in commercial fish culture in rural ponds. This lucrative activity has the effect of making ponds unavailable for some domestic uses, especially bathing with soap. But it is still easy to find women gossiping as they do their laundry and dishwashing or bathe their young children at ponds throughout the countryside. Fish culture also makes pond water undrinkable.

The fourth trend is the filling-in of ponds and canals. There is limited land available for living spaces in most regions of Bangladesh, and land values are increasing. As the need for new land rises with the expanding population, filling in these man-made water bodies is one way for owners to get it. This trend, of course, further reduces the amount of water available for domestic uses.*

---

lated with place-based practices." Bakker's (2012:30) argument continues, referring to struggles in India and elsewhere over control of groundwater resources: "The real 'water crisis' arises from socially produced scarcity, in which a short-term logic of economic growth, twinned with the rise of corporate power (and in particular water multinationals) has 'converted abundance into scarcity'." (citing Shiva 2002)

*Water resource management in Bangladesh has a positive side. Although some individuals do grab control of water bodies for their own profit, whole villages, and even larger social units often cooperate to develop and maintain water resources. Such coordinated community action can effectively prevent flood or water-logging and enhance irrigation opportunities (Duyne 1998, 2004). A promising approach to rural supply is being promoted by the World Bank: namely, the creation of user-funded, small-scale, piped supply systems that use tube well water drawn from very deep aquifers. This program is managed by the Ministry of Local Government, Rural Development and Cooperatives through a special agency, the Bangladesh Water Supply Programme Project (BWSPP), which is the successor to the terminated Bangladesh Arsenic Mitigation and Water Supply Project (BAMWSP), which closed in 2006.

Photo 1-4. The Atrai River is reduced to a small set of pools during the dry season. Bera Subdistrict, June 2009 (Photo credit: Suzanne Hanchett)

Photo 1-5. Women and children bathing in a pond: A common sight in the study areas (Photo credit: Shireen Akhter)

The Arsenic Problem

The government's official definition of water safety has changed drastically since the late 1990s in Bangladesh. From the 1960s onward, governmental, international and local development organizations had promoted the use of groundwater accessed by hand-pumped tube wells. This campaign eventually succeeded in persuading most of the rural population that "tube well water is safe water," as a common slogan put it. Indeed, it was safe in the sense of being relatively free of pathogens that cause water-borne diseases such as diarrhea, dysentery, cholera, or typhoid—diseases often fatal to young children. The official message changed abruptly, however, around 2000-2001, when high levels of arsenic were found in tube well water of 61 of the 64 Bangladesh districts.*

Arsenic is a naturally occurring toxin found in older alluvial soils of this region. It comes mainly from aquifers located 10 to 70 meters (33 to 230 feet) below the earth's surface. It has no color, taste or smell. Even low-level exposure over a long period of time is considered dangerous. Health risks include lung and bladder cancer, neuropathy, and skin lesions possibly leading to cancer. (Smith *et al.* 2000)

In some of the areas where we have conducted our studies, 80 percent or more of the tube wells produce water from aquifers which are less than 70 meters below the surface. High levels of arsenic were found by one survey in almost 25 percent of the drinking water drawn from tube wells. Tube wells, however, still are widely preferred over the more traditional dug wells and rain-fed ponds as sources of drinking water.

When the arsenic situation first came to the attention of the general public, it caused excitement at all levels of society,

*In neighboring districts of West Bengal, India, the arsenic problem was officially recognized in the 1980s.

including donors and policy makers. Large amounts of aid money flooded organizations willing to tackle the problem, but solutions were elusive. Newspapers issued frequent reports on the health and social problems of people afflicted with "arsenicosis," showing skin discoloration or lesions and neuropathy that could lead to leprosy-like symptoms. The public was generally confused by the news that their convenient and tasty tube well water might somehow be "poisonous." Wealthier, more educated people were better informed than those who were poorer and less well educated. In addition to shock, there was some cynicism. (Hanchett *et al.* 2000, 2002, 2006; Asian Development Bank 2003)

Between 1999 and 2003, hundreds of people paid through various arsenic mitigation projects rushed around the country testing tube well water and painting tube well pumps red if the arsenic content of the water exceeded 0.05 milligrams per liter (or 50 parts per billion), the Government of Bangladesh's official limit.* If the water was below this limit, the pumps were painted green. (Photos 1-6 and 1-7) Water testers informed the public about the arsenic problem in most (but not all) cases. Screening exams identified some 13,000 people in Bangladesh with skin lesions that possibly indicated arsenic poisoning. The arsenic problem is estimated to affect 29 million Bangladeshis and another 7 million in West Bengal, India. (Ahmed 2003:12,14)

Some arsenic mitigation projects have installed safe water options, such as deeper tube wells, or provided household and community filters, but there is no ongoing governmental program or service in Bangladesh to deal with the problem.

---

*The World Health Organization's officially arsenic-safe limit is 0.01mg./L, or 10 parts per billion.

Photo 1-6. If water was found to have a "safe" level of arsenic content, the tube well spout or whole well head was painted green. (Photo credit: Cindy Geers, 18 District Towns Project, ca. 1998-99)

Photo 1-7. Tube well heads were painted red to indicate high arsenic content in the water. (Photo credit: Cindy Geers, 18 District Towns Project)

Researching Water Culture

Information for this book was collected while the team or certain team members conducted fieldwork on behalf of several different water projects or water and sanitation programs in Bangladesh. The goals of these programs were generally acceptable to the intended beneficiaries: reduction of diarrheal disease, promotion of latrine use and hand washing, and removal of arsenic from drinking water. So there were few situations in which cultural norms were violated by project staff. We did identify some important misunderstandings and unexpected outcomes, however.

Our collective fieldwork efforts have gone on for more than 12 years. We are drawing primarily from our field notes on studies done in 18 different subdistricts or towns in eight different districts of Bangladesh between 1997 and 2009 plus a short visit to West Bengal in 2004. We also have done some research among ethnic minority groups, but as four authors are native speakers of Bengali, the Bengali language information is where we feel most confident of the validity of our findings. Our research materials are supplemented by literature review and information on other areas from some helpful colleagues. In 2009, UNICEF Bangladesh funded the team to do some systematic data collection on water cul ture in Comilla and Pabna Districts. Details are summarized in Appendix 1.

Most of our information has been collected using rapid rural appraisal (RRA) research methods. (Rapid appraisal research methods are described and discussed in Chambers 1991 and Kumar 1993.)

Photo 1-8. Kazi Rozana Akhter conducting a focus group discussion. Muradnagar Subdistrict, 2006

The typical approach is for two or more researchers to visit a place for a period of one to 14 days. While there, the team develops an overview of the local population's size and socioeconomic characteristics and the distribution of settlements, conducts interviews, and makes systematic observations. Daily team meetings are important to ensure information-sharing and fact-checking. Many of our studies have included questionnaire surveys of randomly sampled households, but almost all of the information presented in this book was gathered using qualitative research methods, such as focus group discussions, social mapping, body mapping, key informant interviews, and structured observations. Case studies and situation analysis are also part of a typical RRA study.

We always met with both men and women, with poor, rich, and middle income people, and with both Hindus and Muslims, if both groups are present in an area. We have attempted to cover places in a broad way, not to neglect remote sections of a visited village.

During the years of research covered by this book, we sought out more than 70 key informants knowledgeable in local lore and practices, and we conducted more than 90 focus groups and other less formal group discussions during village visits ranging from two hours to two weeks. The management of water inside the homestead is almost always the responsibility of women, so women's voices prevail here. Approximately a third of our interviews have included men, so their views and practices are represented as well.

These research methods have both advantages and disadvantages. The main advantage is their efficiency. Group discussions bring out points of community consensus and disagreement on certain issues within a short period of time. Conducting such discussions, we use general questions that will stimulate discussion. Participants are made to understand that we do not expect there to be right or wrong answers. Rather, we are interested to know how people think. Because Bengali villagers generally enjoy discussion and debate, the method brings out plenty of opinions and clarifies points of consensus or disagreement. The advantage is breadth of information, both geographical and social.

A disadvantage of this approach is that it does not provide much information about how cultural principles and social values play out in the daily life of any one place. The presence of visitors tends to put people on their good behavior. They show their best side to the extent that they are able. We have heard many reports of conflict and observed some, too, but the RRA method does not allow for deep exploration of specific situations.

Overview of the Book

Chapter 2 reviews water-related mythology, traditions and folklore in the South Asian subcontinent, and ethnographic studies relating to various types of water. Chapter 3 presents Bengali myths, legends and sayings relating to water generally, to rain, and to various kinds of water bodies. Perceptions of water's qualities and some water categories and Bengali terms are discussed in Chapter 4. A more complete list of water-related vocabulary is in Appendix 2. Detailed information on management of domestic supply and uses of water in healing and family rituals is in Chapter 5. In Chapter 6, we return to the arsenic problem, efforts to solve it, and culturally based reactions to it. Chapter 7 summarizes some principles of water culture in the study areas.

2. Water in South Asian Traditions

Numerous philosophical and religious movements—Hindu, Buddhist, Jain, Muslim, and others—have formed the cultural history of South Asia for more than 3000 years. Every region and religion has been affected by multiple influences. And every country, state, and district has its own complex mix of beliefs and practices. Nonetheless, there are numerous common patterns and themes in the traditions of different places and peoples. Although the population of Bangladesh is almost 90 percent Muslim, much popular or folk culture is shared with Hindus, Christians, and Buddhists. A brief review of South Asia's water-related traditions can shed light on the varied historical sources of some widespread folk customs and cultural precepts.

Early Influences on South Asian Water Culture

An early civilization along the Indus River in what is now Pakistan devxeloped extensive water-works between the fourth and third millennia BCE (Before the Common Era, sometimes written as BC). The ancient city of Mohenjo Daro had a network of covered drainage channels and rooms with drains in their floors. There also were many large bathing areas, suggesting that water was used for purification. But there are no written records documenting water-related beliefs in the Indus Valley Civilization.

Archeologists have found massive stone dams for water storage in multiple locations outside of the cities. Drains and wells existed even in villages. (Fairservis 1979a, 1979b) Ever since this early period, elaborate water resource management systems have been part of South Asian social, cultural and economic life*.

In the middle of the second millennium BCE, Aryan peoples —speakers of an Indo-European language—entered South Asia from the northwest. Unlike the early Indus Valley people, they did leave behind texts, the *Vedas*, which are the earliest documented basis of Hindu beliefs, myths, and rituals.**

From these ancient times through the medieval period, religious literature devoted much attention to water. According to Radha Krishnamurthy,

> In ancient India water was used in all religious rituals and ceremonies because it was believed that the pure, divine, well-provided waters convey the offerings to gods. Water, though itself a purifying agent, was held to be very sacred and people were often exhorted not to harm waters which are full of saps and good food. ... Water plays an essential role in [humanity's]...physical and mental development. (1996:327-328)

Such ideas continue to influence beliefs and practices of Hindus, and are evident in the life ways of other South Asian communities as well.

---

*In Sri Lanka and throughout southern India large water storage tanks were built by numerous early kings. Water control structures tapped the Kala Oya in the 5th century BCE, diverting river waters to supplement the supply of water to the capital city, Anuradhapura, and its surrounding area. (Disanayaka 2000:16)

**In its form at the time of its final edition, the earliest Aryan document, the *Rigveda*, reflected a well-developed religious system. The date commonly given for the final recension of the *Rigveda* is 1200 BCE. (*Encyclopaedia Britannica* online, "Hinduism", 2012)

Historical Influences on Bengali Water Culture

The Bengali-speaking region saw Buddhism flourish from the third century BCE as Mauryan emperors extended their influence. From that time until the 12th or 13th century CE (Common Era, or AD), Buddhism and Hinduism coexisted as a succession of different dynasties controlled the region.* From the 14th century CE onward, Islam has been a dominant influence, especially in the eastern part of the Bengal region, most of which is now Bangladesh. Before the advent of Islam, however, Hinduism was well established in the western parts of Bengal, where rice culture supported a Brahmin religious aristocracy and Hindu temples. In eastern Bengal, according to Ira Lapidus, Brahmin and caste influences were weaker, and Buddhism was the dominant religion. "Muslim rule put East and West Bengal under a single regime permeated by Islamic symbols....Agricultural expansion into East Bengal made converts out of newly sedentarized peoples." (Lapidus 2012:513-14)

Islam had been introduced to the Indian subcontinent through a series of Arab invasions starting in Afghanistan in the eighth century CE, but "The conquests of India began in earnest...in 1030," when a group known as the Ghaznavids (or Ghaznavi) captured Lahore and plundered North India. (Lapidus 2012:509,514 *et passim*) By the 13th century (1236), the Delhi Sultanate had extended its reach eastward into the Bengali speaking region, and by 1335 it controlled most of the territory that is now Pakistan, India, and Bangladesh.

Islamic conquerors in Bengal, as elsewhere in South Asia, encountered a fully developed set of Hindu and Buddhist cultural traditions. Eventually Arab, Turkish and Persian elements were added to these traditions. As more and more people

---

**Encyclopaedia Britannica* online, "Buddhist, Hindu, and Muslim Dynasties until c. 1700"

converted to Islam, social beliefs and practices blended old and new ways. Different Muslim rulers, it must be noted, had different approaches to the mixing of cultural traditions. According to Lapidus, those controlling Bengal territories ruled with a culturally light touch. "...While the Delhi sultanate leaned to Muslim supremacy," Lapidus says, "the provincial Muslim regimes fostered the integration of Muslim and Hindu cultures and the formation of an Indian version of Islamic civilization." (2012:513)

> [In] popular culture the boundaries between Islam and Hinduism were more flexible than in formal doctrine.... In Bengal and the Punjab Muslims celebrated Hindu festivals, worshiped at Hindu shrines, offered gifts to Hindu gods and goddesses, and celebrated marriages in Hindu fashion. Hindus who converted to Islam retained many of their [past] beliefs and practices; many Hindus venerated Muslim saints without a change of religious identity. (Lapidus 2012:515)*

This historical blending of people and their varied traditions makes distinguishing influences on Hindu *vs.* Muslim folk practices difficult, although formal religious rules differ. For many generations, these two populations have lived as neighbors and shared basic cultural precepts. In our study of Bangladesh village water culture, we found only a few points of difference between the two groups' water-related ideas and practices.**

---

*According to Lapidus, unlike the Arab or Turkish Muslim conquerors of the Middle East, those who invaded South Asia were not backed by large numbers of settlers. "The Muslims established a political elite, but they could not colonize the country." Furthermore, "the pre-Islamic political structure of India remained intact. The previous regime was highly decentralized, composed of local lords and a Brahmin religious elite who retained local political power under Muslim suzerainty.... As in the Balkans, the thinness of the governing elite, the confirmation of local lords, and the protection of non-Muslim populations served to maintain the continuity of non-Muslim identities." (2012:516)

**Unfortunately, these cultural commonalities have not prevented sectarian tension and conflict.

Water, the Cosmos, and the Human Body

Early Hindu religious texts envisioned creation as starting from a cosmic ocean that surrounded the universe: "...the earth, the upper and infernal regions, and all their beings [having been] shaped out of the cosmic waters of the abyss." (Zimmer 1972:62) Vedic texts declared that ceremonial use of water could purify the body and spirit, neutralizing the effects of sin or evil, which were said to cause "illness, enmity, distress or malediction."*

Early philosopher-scientists in India, Babylonia, Greece, Egypt and China recognized water (called *jala* or *ap* in India) as a basic cosmic element. Other elements recognized in early Indian texts were fire (*tejas, agni*), earth (*bhumi, kshiti*), air or wind (*marut, pavan, vaayu*), and ether (*aakaash*). A. L. Basham explains that,

> Ancient Indian ideas of physics were closely linked with religion and theology, and differed somewhat from sect to sect. As early as the time of the Buddha, if not before, the universe was classified by elements, of which all schools admitted at least four—earth air, fire and water. To these orthodox Hindu schools and Jainism added a fifth, ākāsha, which is generally translated 'ether'. (Basham 1959:496-497)

Sanskritic literature included elaborate classifications of water and discussions of its qualities. These are described in some detail in Radha Krishnamurthy's 1996 article on "Water in Ancient India." Water was broadly classified into two types: that which falls from the sky (*divya*) and surface or groundwater (*bhauma*). Each of these types was further subdivided. Water from the sky was either continuous rain, hail, snow or dew. Two types of rain

* *Encyclopaedia Britannica* online, "Hinduism" (2012)

were distinguished depending on the season in which they fall. Nine types of surface or groundwater were described: including, among others, the "water of rivers emerging out from mountains and flowing in the fertile regions," and "the clear and tasty water with the hue of blue lilies collected from ponds and wells." Rain water was said to have six qualities: "coldness, purity, benevolence, pleasantness, clearness and softness."* Other properties, such as flavor or alkalinity, "are acquired by water after it falls on the ground," as it is affected by the qualities of soil or environment and seasonality. The various types of water were thought to affect the human body in different ways:

> Thus great physicians and seers of ancient India were aware of the different properties of rain water in different seasons." In the rainy season, water was said to be "heavy and greasy." During the autumn it is "[thin/light] and non-greasy." Water in winter is "unctuous, aphrodisiac, strength-promoting and heavy." (Krishnamurthy 1996:332-334)

Krishnamurthy's review of texts finds mention of six types of water pollution, a test to determine whether water is safe for consumption, and suggestions about water purification.** Some of these concepts resemble those we have encountered in our Bangladesh studies (see Chapter 4).

Islamic philosophers over the centuries also devoted attention to water, but their emphasis, according to available literature, is on the human responsibility to properly care for this gift from God. Sharing water is a good deed, and not sharing is a sin. (Faruqui *et al.* 2001)

---

*Krishnamurthy's source is *Caraka, sutra* 198. (Krishnamurthy 1996:331)

**Rain water is said to be either contaminated (*sāmudra*) or not contaminated (*gāṅga*), depending on the season and other factors. The type can be determined by whether or not it changes the color of a lump of cooked rice placed in a silver vessel. "If the rice changes colour, the water should be taken as sāmudram which is not fit for use except in the month āśvina." (Krishnamurthy 1996:331)

Water plays an essential role in Islamic spiritual life. Without washing in water, one cannot join in the daily prayers. Both men and women are prohibited from addressing God if the body is unwashed. Head-to-foot washing with water is required of men and women wishing to perform the rites of pilgrimages to Mecca and other holy sites.

Muslim pilgrims frequently bring back holy water from a well in Mecca called either Jamjam or Jumjum, in hopes that it will help to fulfill wishes, cure illness, succeed in childbirth, and otherwise help solve life's problems. The story, according to one source, relates to Hajera's frantic search for water to keep her and her son Ishmael alive after they had been cast out by Abraham. She eventually found a well. "The gushing water from the well was making all that sound. She quickly puts a barrier around it so that water could be stored and that it did not overflow. And this well came to be known as the Jamjam."*

---

*Some Bengali Muslim villagers refer to human sexual fluids by the term "impure water" (*naapaak paani*), because semen resembles water in its color, liquidity and other characteristics. The healer-philosophers, or* fokirs, *say that, "God creates humans by this profane water." They explain that the first couple, Adam and Howyaa [Adam and Eve], had no reproductive power until they ate fruit that God forbade to them. This fruit produced semen in Adam's and Howya's bodies. The semen was "pure" (*paak*) at first, but when the couple enjoyed sex with each other, it was transformed into "impure water" that produced children.*

Comments of a Muslim man in
Bhuapur Subdistrict about fokir wisdom

---

*Siddiqui n.d.

People who bring back Jamjam water distribute it to many others, who mix it with normal waters to purify them. Some uses in childbirth are discussed in Chapter 5.

The practice of carrying back water from a sacred site is observed by Hindu pilgrims traversing the South Asian countryside as well. Diana Eck's study, *India: A Sacred Geography* and Anne Feldhaus's Maharashtra study, *Connected Places*, both describe this practice and the ideas on which it is based. Eck describes the town of Hardvar as "inundated with pilgrims carrying *kavads*—poles slung over the shoulder with water pots on each end," during the winter month of Phālguna (Faalgun) or the summer month of Sravana. The pilgrims come to this place from throughout the region, "fill their water pots, and return to their towns and villages with Gangā [Ganges River] water to pour upon the images of Shiva in their own locale." Other uses for water brought from pilgrimage sites include healing and purification. (Eck 2012:144)

A vivid, personal account of the Hardvar event by the environmental activist, Dr. Vandana Shiva, reveals the strong emotional ties that people have to their sacred river:

> A few years ago, a few thousand pilgrims used to walk from villages across north India to Hardwar and Gangotri to collect Ganges water for Shivratri, the birthday of the god Shiva. Carrying *kavads* (yokes from which two jars of holy water dangle and are never allowed to touch the ground) the *kavadias* now number in the millions. The highway from Delhi to my hometown, Dehra Dun, is shut during the weeks of the pilgrimage. Villages and towns put up free resting and eating places along the entire 200-kilometer pilgrimage route. The brightly decorated *kavads* containing Gangā water are a celebration of and dedication to the sacred. (Shiva 2002:138)

Water and Health: the Basis of Hot-Cold Thinking

According to Ayurvedic medical theory, like that of other humoral systems, "Illness is due to upsetting the homoeostatic condition of the *tridoṣa* [humors]." (Obeysekere 1976:202) Balance among the humors often is spoken of as adjusting the heating effects of the element fire against the cooling effects of the element water. Illnesses often are understood as excesses of cold or heat; and dietary practices and herbal medicines are meant to restore health by restoring balance among the humors.

Many foods, herbs, and other items are classified as heating or cooling in terms of their physiological effects. In humoral theory hot and cold qualities are distinct from measurable temperature. As we will discuss in Chapters 4 and 5, folk ideas about water still are influenced by this ancient theory.

The Bengali speaking region of South Asia, and especially Bangladesh, has been influenced by two schools of humoral medicine, Ayurveda and Yunani (or Unani). These two medical traditions have different origins. Ayurveda is indigenous to South Asia and based on early Hindu texts and Brahmanical traditions dating from the first millennium BCE or earlier. Yunani is based on early Greek texts, which were translated into Arabic by Muslim scholars. Yunani was carried by the spread of Islam into Central Asia, India, and Southeast Asia from the eighth or ninth century CE onward. (Leslie 1976:1-3; Bürgel 1976) Practitioners of the two systems were in contact early, but they did not influence each other substantially until the 19th or 20th centuries, according to the anthropologist George Foster.*

*According to Foster, the Yunani/Unani medical system, developed by Muslim scientists from the seventh century onward, is based on Greek cosmology and humoral theories. Christians had made Syriac (Aramaic) translations of Greek medical texts during the fourth to sixth centuries A.D. These were translated into Arabic after the Muslim conquest of southwestern Asia and North Africa, he ex-

Humoral medicine practitioners are still being trained in South Asia. According to medical anthropologists, however, most now combine modern medical treatment techniques with traditional ones rather than following humoral theories in a pure or strict sense. Humoral theories have survived mainly in popular health beliefs and folk healing. According to several ethnographic reports, a typical Bangladesh village has healers who combine ideas and methods of Ayurvedic and Yunani traditions, along with folk ideas of disease causation. Many folk healers are women. Their patient populations are largely female, as men tend to make more use of professionals trained in western medicine and employed in government or other facilities.

Water Mythology

Water has many other values and meanings in the South Asian region beyond those related to health and illness. The story of the descent to earth of the river and goddess, Gangā (pronounced *gaŋgaa*), is the best known, being found in many classic texts, such as the *Purāṇas* and the *Rāmāyana*.* Though the Ganges is the

---

plains, forming the basis of early Arab medical research. "Beginning in the 7th century, Greek humoral medicine, in its Galenic-Persian form, diffused eastward in Asia, carried by the expanding Muslim tide. The Muslims call this medical system Tibb-i-Yunani, or Unani ('Ionian', i.e. Greek), and in the countries in which they settled, it has demonstrated remarkable tenacity both at folk and sophisticated levels." Arabic texts eventually were translated into European languages, introducing humoral theories into medieval European medical knowledge as well. (Foster 1994:12-13) A Wikipedia article claims that, "....Unani medicine is very close to Ayurveda. Both are based on theory of the presence of the elements (in Unani, they are considered to be fire, water, earth and air) in the human body.... According to followers of Unani medicine, these elements are present in different fluids and their balance leads to health and their imbalance leads to illness." (Wikipedia 2012, "Unani," http://en.wikipedia.org/wiki/Unani)

*"My Hindu neighbors believe that Gangaa Devi, the river goddess, lives in water. "Of course, there are gods and goddesses everywhere in the world," explains one woman. 'Gangaa is our mother river. She controls life by giving us water'." - Delduar (Tangail) report by Anwar Islam, 2012

river principally associated with the goddess, different versions of this myth are repeated in relation to different rivers in different geographical settings, according to Diana Eck. She summarizes the story in this way:

> Gangā was originally a divine river, streaming across the heavens....Through the asceticism and prayers of the sage-king Bhagīratha, she agreed to descend from heaven to earth to raise the dead ancestors of the solar kings of Ayodhyā. To break the force of her fall, Gangā fell first upon the head of Lord Shiva in the Himalayas and then flowed across the plains of north India.

Myths associated with other sacred rivers, such as the Godavari and the Narmada, "repeat this pattern of divine descent. (Eck 2012:19)

Themes of water related myths in classical literature, such as the *Vedas* and *Purāṇas*, are quite varied, as Kumar (1983) and Baartmans (1990) have shown. Prominent among these themes is metamorphosis, or water having been "attributed with the power of causing transformation by its mere touch. Many a *tirtha* [holy water-place or pilgrimage site] are eulogised for their supernatural power causing change of form to various beings." Kumar cites examples from the *Skanda Purāṇa*:

> ...Śiva gets attracted by Tilottamā's beauty and Pārvatī [Śiva's consort] curses her to become ugly. But afterwards Pārvatī, finding her innocent, herself takes her to *Rūpadatīrtha* for a bath where she regains her original form. (Kumar 1983:272)

In another tale, ...Raṁbhā was cursed to become a rock by the sage Viśvāmitra. ...She was freed from the curse by the touch of

Photo 2-1. Representation of the Hindu goddess Gangaa as a beautiful woman. She is identified with the Ganges River. (Photo credit: Suzanne Hanchett)

holy water only. Sage Śveta threw this slab (of rock) on a demoness to kill her. Both the demoness as well as the rock fell into *Kapitīrtha* and the holy-water changed the rock back into its form as Raṁbhā. (Kumar 1983: 274)

Gangā is not the only deity or spirit associated with water in folk thinking. In 1894, Crooke wrote about a "god of water" revered by villagers in what is now Uttar Pradesh, India. "The Hindus have a special god of water, Khwája Khizr, whose Muhammadan title has been Hinduised into Rája Kidár, or as he is called in Bengal, Káwaj, or Pír Bhadr....Among Muhammadans a prayer is said to Khwája Khizr at the first shaving of a boy. At marriages a little boat is launched on a river or tank in his honor.... All through the Eastern Punjáb he is intrusted [*sic.*] with the safety of travelers." (Crooke 1894:26-27)

Belief in this spirit persists into present times and extends to Bangladesh villages. The ethnographer Thérèse Blanchet mentions "Kwaz, the guardian of water" as one of several godlings who "usually help enforce a code of conduct which maintains society in a civilized state" in the Jamalpur District (Blanchet 1984:41). We also have heard mention of this spirit in our working areas. Muslims in Delduar Subdistrict generally believe that a spirit named Khoyasj Khigir lives in water to protect this necessity of life.*

Vessels

According to J. B. Disanayaka (1984:8-9, 16), Sri Lanka folklore and ritual—like that of other regions—uses full and empty vessels in poetic and metaphorical ways. A mother is compared to a full vessel, for example, "and a woman who carries an infant is considered lucky because ... her breasts are full of milk." Empty vessels, on the other hand, "fall into the category of unlucky symbols," Disanayaka writes. "Empty vessels... are supposed to bring ill-luck, if they are met at the outset of a journey." Such empty vessels can include barren women or old women. By extension, round things are considered lucky and flat things unlucky. Empty vessels are inauspicious symbols elsewhere in South Asia also.

*Report by Anwar Islam, 2012

The full and round breasts of a young woman are compared to golden pots in a poem that Disanayaka quotes:

Keeping the golden pitcher on the waist
Covering two golden pots with a silken garment
Bedecked with both silver and gold
Whither goes, the Queen of Love, at the break of dawn?

Water vessels are used in many ways in South Asian folk ritual symbolism. Water contained in a vessel, for example, can be used in Hindu ritual to incorporate an otherwise disembodied spirit. An ethnographic report on home-centered rituals in Karnataka State, in southern India, describes pots of water being used to temporarily embody ancestors or other spiritual beings and to receive offerings made to them. (Hanchett 1988) Using water-filled containers in this manner is common Hindu ritual practice in Bengal and other parts of the subcontinent.

Bathing and Purity

Ethnographic studies throughout the South Asian region make frequent mention of bathing and water as an essential part of rites of passage and other folk rituals. According to Zeitlyn and Islam's research in Chandpur District and some urban areas of Bangladesh, "Water is regarded as the agent of purification *par-excellence* by both Hindus and Muslims." "The simple act of pouring over or immersing the body in water is regarded by both groups as purifying without any scrubbing or other actions because of the inherent qualities of the water.... [This] has little to do with germs." (Zeitlyn and Islam 1990)

Baths in West Bengal rites of passage (*saṃskāras*) have special meanings which reflect more widespread South Asian ideas. These meanings highlight important differences with Western ideas of purification by bathing. According to Inden and Nicholas,

> ...There is one important respect in which the *saṃskāras* differ from the [Christian] sacraments: The *saṃskāras* are thought to affect the total person of the recipient and not only a 'spiritual' part. Thus, for example, bathing is regarded as purifying the 'mind' (*man, manas*) and 'heart' (*hṛdaya*) as well as to the exterior surfaces of the body. The head and heart are believed to control a person's body. Sprinkling the head with water and touching the region of the heart in the *saṃskāras* are conceived of as actions that affect what Westerners would regard as both the 'material' and 'spiritual,' or 'physical' and 'mental' parts of the body. (1977:37-38)

These social and spiritual observations help us to understand the special meanings of "purity" in the South Asian context. At the most general social level, the Hindu caste system depends on this idea. The caste system keeps Dalits (formerly spoken of as untouchables) at the bottom and Brahmans at the top. Caste ideology requires people to marry within their specific caste or subcaste group.

As Louis Dumont has explained, each caste strives to maintaini its separateness and its "purity." In the case of Brahmans, for example, caste purity "is acquired by generations of pure conduct, which consists of doing actions that are pure, eating pure food, by increasing his own personal sacredness, by the study of the Vedas, and by marriage only with people who have kept pure conduct..." This ideal of "interior purity," among other things, governs the separation of one caste from another. (Dumont 1972:326)

Water Sharing and Social Hierarchy

In the Hindu caste system, rank is expressed partly through rules of water exchange: that is, which group normally gives or takes water from which other groups. Castes whose members share water (or foods cooked in water, such as cooked rice) are defined as equals. Higher-ranked castes give water but do not take it from lower-ranked castes. The touch of a lower caste person is thought to make a water-containing vessel polluted and thus unacceptable to people of higher castes. Giving, taking or sharing of water thus is a sensitive matter in Hindu life. It also affects the sharing of village water collection places (water points) and other social contacts. A 19th century example provided by Risley illustrates the limitations on water-sharing between castes, limitations still observed, especially among Hindus:

> The Khatya claim to belong to the Maghaiyá potter family of Patna. They drink water from the vessels of the other Kumhárs [Potters], and may give water tothem, but hold no communication with the Rájmaháliá Kumhárs. None of the other Bengali Sudras, however, admit their equality.... One sub-caste of Potters in Pabna is Sirasthan. They are believed to have come from the North-Western Provinces....Their habits are supposed to be unclean, and Brahmans will not take water from their hands." (Risley 1892:518-519)

For Muslims, the hierarchical implications of taking water are important, but this act does not have the same caste meaning that it has to Hindus. An important exception is made, however, when it comes to contact with so-called "Sweepers," whose hereditary caste work is to clean pit latrines and other defecation places. While studying sanitation practices in 2009-2010,

Photo 2-2. Pond next to a village mosque, where men are purifying themselves before prayers. Comilla District, Muradnagar Subdistrict, 2006 (Photo credit: Suzanne Hanchett)

we interviewed several members of the Hindu Methor (Sweeper) caste who told us with some bitterness that they are not allowed to take tea at common tea stalls because their touch is considered—by mostly Muslim patrons and shopkeepers—to pollute the tea cups and glasses. Even when they go to market to purchase food items such as vegetables, pulses or rice, they cannot touch them directly. Rather, they have to lay a piece of cloth or towel on the ground to receive items they purchase from the shopkeeper.*

In her Jamalpur ethnographic study, Thérèse Blanchet discusses the low-caste status of the Sandar people, an itinerant Muslim group:

*Kazi Rozana Akhter, Sweeper interviews from Lakshmipur, Manda, and Naogaon. In India, the Sweeper castes and others formerly called "Untouchables" have decided on the group name of Dalits.

> [Their] status in Bangali Muslim society is certainly interesting. Locally they are not refused the title of Muslim but they are considered as an inferior caste. No one will eat with a Sandar, no one will take water from them...." (Blanchet 1984:104)

Human feces are considered to be one of the most polluting substances in this spiritual sense. Contact with feces, therefore, is used to justify the low status of Sweeper castes. We met with a number of Muslims who had taken up the toilet-cleaning business. Most of them were working in a different district, far from their homes, and keeping the source of their livelihood a secret from their relatives. One male Muslim latrine-cleaner in Chittagong District speculated that, "Even if we do it secretly, maybe one day they will know our profession and send back our [married] daughters to us. Now we are thinking to arrange marriages within the 'Muslim Methor' community." Their new occupation, they said, was likely to force them to become a new caste.*

Unlike taking it, *giving* water to anyone who is thirsty is promoted by Bangladeshi Muslim villagers as a great virtue. This is regarded as a life-saving gesture and an obligation of a good person, one that will result reward the giver in Heaven. An imam (mosque-based Muslim religious leader) of Delduar Subdistrict in Tangail told us the following story:

> *"Once there was a fight between two groups in Mecca. One was Muslim, and the other, a non-Muslim. Many injured fighters were crying out, 'I am thirsty. Please give me some water, so I can survive'. A Muslim woman approached the two injured men and offered some water to the Muslim fighter. He looked at the water and started to take a drink from the pot, until he remembered the*

*Shireen Akhter field report, 2010. See Hanchett 2011 for further discussion of this situation.

*saying: 'Give water to your enemy first if he wants it'. (*Charom shotru keo paani daau.*) Suddenly he rejected the water and asked the woman to give it first to his enemy, who was more badly injured than he was. The woman went to the non-Muslim and offered him water. He rejected her offer, repeatedly requesting his Muslim enemy to drink first. At that point, both of them died."**

Ethnographic Studies of Water and Culture in South Asia

A few scholarly studies have described and discussed social and cultural aspects of water life in the South Asian sub-continent. The most detailed report is by Bernadette Maria Gomes (2005), who emphasizes the many ways in which water serves as "the eco-cultural reality" in Goa. She declares that in Goa,

> Water has generated communities and fostered a sense of community, charted the course of history, evolved beliefs and ritual, contributed to oral traditions of proverbs and idioms, and it remains the matrix of social life for the Goan people.

Water rises above being a mere resource, according to Gomes. The varied social and cultural values attributed to water, she argues, "...form a part of [the Goan] quotidian ethos. The folk conception of water as a physical substance is *udok*. As a spiritual entity, it is *tirth* or *azmente* having a character of its own."**

*Anwar Islam report, 2012

**Consecrated water is called *tirth* by Hindus, "and the Portuguese term *azmente* is used by the Christians in Goa." [Gomes 1983:270] *Udok* means "water" in the Konkani language.

Spring waters and sea water both are popularly believed to restore health. Like many Bengalis, people of Goa make ample use of fish in their diet.

> Fish is the most important food item obtained from the waters. Popular literature has often rightfully termed fish as the soul of Goan cuisine. Even the socially powerful and ritually pure Saraswat Brahmins of Goa have traditionally been 'fish-eating Brahmins'. (Gomes 2005:265-270)

Water Language

Language is the primary means by which humans create and communicate ideas about water, or any other aspects of the environment. J.B. Disanayaka reports that Sinhalese sea fishermen use special words in talking about rain when they are at sea, because at sea "one is…completely dependent on the protection of gods, [so] profane words are avoided." (Disanayaka 2000:131)

Erlend Eidsvik finds that different Nepali words for water have different connotations. "Briefly, *jal* means holy, auspicious, sacred water, while *pani [paani]* is the neutral lexical denotation of water. [Etymologically], *pani* means 'drinkable'.... As one informant [says], 'I agree that Bagmati [a sacred river] is *jal*. All the other water is just water, *pani*." (Eidsvik 2003:6)

In Bangladesh, Muslims customarily use the term *paani* for water, and Hindus call it *jal*. Issues of water-related language will be discussed further in Chapter 4.

Rain

Numerous South Asian myths and associated customs relate to rain: bringing it, stopping it, and otherwise controlling it. Some legends refer to sacrifice as necessary for production of rain or other water sources. Professor J. B. Disanayaka (2000) describes a popular Sri Lanka rain idea:

> Our sacred [*Ficus*] tree, *boddhya*—we believe that if you pour water into that tree during drought, it will bring rain....The image of the Buddha...can also call rain.... For three months [Buddhist] monks go into 'rain retreat'.

Reports from Bengali and Assamese areas depict rain-making ceremonies that make magical use of feminine erotic power to bring rain.

> In some areas of Bankura and Burdwan districts of West Bengal....the entire womenfolk of the village, irrespective of caste and creed, go to the field headed by an elderly Brahmin woman. [They] carry all the implements that are generally used by the farmers, namely spades, bailing implements, hooka... etc. In the field they worship [several gods].... An immature married girl... [lies] on the field naked and two others spray water on her thrice.... In Assam the Koch women [plough a field] in a naked state... to compel rains." (Mahapatra 1963:64-65)

Crooke's 19th century research on north Indian folklore turned up similar practices outside of the Bengali-speaking region.

> [During the Gorakhpur* famine of 1873-74] there were

*Gorakhpur is a city located in what is now eastern Uttar Pradesh, India.

> many accounts of women going about with a plough at night, stripping themselves naked and dragging it across the fields as an invocation to the rain god….In Chhatarpur, when rain falls, a woman and her husband's sister take off all their clothes and drop seven cakes of cow dung into a mud reservoir for storing grain." (1894:41)

Crooke quotes a dramatic and supposedly successful rain-making ritual described in *North Indian Notes and Queries* by Bhān Pratāp Tiwāri in 1891. Of the regular nudity spell in case of failure of rain, we have a good instance from Chunár in the Mirzapur district," Crooke wrote:*

> The rains this year held off for a long time and last night (24th July 1892) the following ceremony was performed secretly. Between the hours of 9 and 10 p.m., a barber's wife went from door to door and invited the women to join in ploughing. They all collected in a field from which all males were excluded. Three women from a cultivator's family stripped off all their clothes: two were yoked to a plough like oxen, and a third held the handle. They then began to imitate the operation of ploughing. The woman who had the plough in her hand shouted, 'O Mother Earth! Bring parched grain, water and chaff. Our stomachs are breaking to pieces from hunger and thirst'. Then the land-lord and village accountant approached them and laid down some grain, water and chaff in the field. The women then dressed and returned home. By the grace of God the weather changed almost immediately, and we had a good shower." (Crooke 1894:43)**

*Chunar is a city in Mirzapur District, Uttar Pradesh, India.

***North Indian Notes and Queries*, vol. I, p. 210 (Allahabad: Pioneer Press). This volume has been digitized by Google.

Photo 2-3. Folk art showing Behula on a banana-trunk raft with her dead husband on her lap, traveling along a river in search of a way to bring him back to life (Photo credit: Dalbéra Jean-Pierre, via Wikimedia, commons.wikimedia.org)

P.K. Maity (1988:29-39) describes some Bengali folk rituals conducted by women to bring rain. Called *purṇipukur brata*, these ceremonies follow the principle of sympathetic magic, or like-affecting-like. A group of women together dig a small hole, which they call "pond" (*pukur*), at a good location and decorate it with ritually important plants and flowers. They then carry some items usually used for worship along with a small brass drinking pot full of water and sit near the little "pond" reciting special verses as they pour water into it and on the plants inside it. This ritual is performed by married women or maidens. Maity reports that such rituals are less frequently performed nowadays than in earlier times "due to the advancement of science." (1988:38)

Rivers and the Sea

Rivers are prominent in South Asian literature, folklore, and religious life. Waterways are a prominent feature of many parts of what the religion scholar Diana Eck (2012) calls "the storied landscape" of South Asia. Ancient myths and contemporary folklore alike personalize and deify rivers. The most well-known of these, of course, is Gangā.

> ...Indian civilization, preserving [an] emphasis on running water and purification, has developed a full range of mythological and ritual traditions concerning sacred rivers....The waters of the Ganges, identified with the milk of mother cows, are truly life-giving waters, and are called 'mother' as they are sipped by devout Hindus. (Eck 1987:426)
> ....The sea is often described as the 'lord' or 'husband' of the rivers. The rivers are, on the whole, female, and Gangā describes the 'womanly character of rivers' in the Rāmāyana....The rushing of a river toward the sea is

> often described as the rushing of a woman in love toward her lover. The meeting of the rivers and the sea is, therefore, like an auspicious marriage. (Eck 2012:151)

In her study of Maharashtra State pilgrimage sites, Anne Feldhaus finds that in popular religious literature, "the union of a river and the ocean presents a powerful image of fulfillment: of reaching a goal, on the part of the river, and of replenishment, on the part of the ocean," whether the union is seen as a marriage or not. (Feldhaus 2003:22)*

The place where two or more rivers merge is called *sangam* in Bengali and other South Asian languages.** Such places often are considered magical and holy. They may be sites of special rituals or pilgrimage. Their magical quality can be at least partly explained by a common folkloric principle that the merging of separate or opposed things produces a special sort of power or energy. The place where roads cross may also be viewed in this way. Such merging or crossing produces an ambiguous state or entity, and symbolism in folklore typically endows structural ambiguity with unique powers.

Feldhaus uses Maharashtra State interviews and textual analysis to demonstrate that fecundity is the "central meaning of rivers in India in general." Rivers are considered good places to give alms and perform sacrifices, she argues, because they "embody the natural world's generosity to humans and thus inspire humans to perform rites of generosity of their own." (Feldhaus 1995:186; 79)

Kuntala Lahiri-Dutt uses Bengali literary sources to show that

---

*Feldhaus, a religion scholar, makes use of documents called Māhātmyas. "A Māhātmya is a type of traditional verse text that glorifies something. Generally a Māhātmya glorifies a particular holy place, but there are also Māhātmyas that praise a particular ritual practice, a certain month of the religious year, or … a river." (Feldhaus 2003:18-19)

**This word in Bengali also means "copulation."

rivers often are represented as feminine powers with personalities and moods. The feminine metaphors used to represent rivers are a powerful mix of creative and destructive forces (She notes the irony in this view, considering that actual females in India tend to be socially powerless.). A quotation from Kabir's *Men and Rivers*, set in the difficult living environment of the sandbar island (*char*), illustrates the ways that "feminine" rivers can embody both creative and destructive qualities:

> Nazu Mia stood on the bank of the Padma [River] and looked around. Before him stretched the waters of the mighty river. In the morning light she seemed placid and content. There was no suggestion now of her power and cruelty. (Kabir 1947, quoted in Lahiri-Dutt 2006:400)

Crooke's 19th century folklore studies showed that "all rivers are not beneficent." He described one stream, the Karamnása, that has an "evil reputation" among Hindus:

> The legend of this ill-omened river is connected with the wicked King Trisanku....When the sage Viswámitra collected water from all the sacred streams of the world, it fell burdened with the monarch's sins into the Karamnása, and has remained defiled ever since....Even nowadays no good Hindu will touch or drink it, and at its fords many low caste people make their living by conveying on their shoulders their more orthodox and scrupulous brethren across the hated river. (Crooke 1894:22-23)

Rivers are generally considered to be purifying agents, but some priestly lore claims that the rivers themselves periodically need purification. In her study of Hindu pilgrimage sites, Diana Eck describes such an occasion at a month-long festival in

January/February in the floodplain of the Triveni Sangam at Allahabad, Uttar Pradesh.*

Kelly Alley (2002) worked in four cities along the Ganges River, where she analyzed discourse (among government officials, political leaders, religious specialists, environmental activists, and others) about the effects of waste dumping on the sacred river. Some religious leaders and ritual specialists insisted that "river pollution is an altogether erroneous notion, and Gaṅgā's powerful flow and motherliness attest to its fallacy....Scientific treatments to 'clean' the river...cannot reach or transform divine power." (Alley 2002:220) Vandana Shiva cites research showing that "cholera germs die in Ganges water." (2002:134) Alley's and other studies highlight a challenging (to environmentalists) gap between traditional thinking and environmentalism.

Lahiri-Dutt and Samanta, in their study of delta sandbar island (*char*) dwellers, argue that, "Rivers are always present in the collective consciousness of Bengal, not just as symbols or even as physical manifestations of cultural meanings, but forming the relations between places." Rivers, they find, are part of communities' "self-definition," an important "element in people's sense of place." (2013:201)

In the southern part of West Bengal, where the Hoogly River, an arm of the Ganges, enters the Bay of Bengal, there is a pilgrimage site called Gangā Sāgara. This is the site of an annual festival. About 80 miles south of Kolkata, this place is associated with the myth of the sage Kapila Muni, who was rudely accused of stealing a sacrificial horse by King Sagara's sons:

---

*Triveni Sangam-Prayag (Allahabad) is the place where the Yamuna River merges into the Ganges. A third, mythical river, the Saraswati, is said to join also, making the site a powerful blending of three rivers. This is the site of the Kumbh Mela, a festival held every 12 years. "The three rivers maintain their identity and are visibly different as they merge. While the Yamuna is deep but calm and greenish in colour, the Ganga is shallow, but forceful and clear. The Saraswati remains hidden, but the faithful believe that she makes her presence felt underwater." (http://en.wikipedia.org/wiki/Triveni_Sangam)

> It was there that he burned them to ash with the fire of his gaze, promising to restore them to life only when the River of Heaven [the Ganges/Gangā] would flow upon the earth. When King Bhagīratha brought the Gangā to earth, he led her across the plains of north India, blowing his conch all the way, finally reaching the sea here at Gangā Sāgara. (Eck 2012:150)

A river is a place of dramatic action and regional connections in the well-known Bengali tale, "Behulā and Lakhindar." * This story features the river as the heroine's path to redemption. The heroine, Behulā, travels on a banana-trunk raft down a long river for six months seeking divine help to bring her dead husband, Lakhindar, back to life.**

Lakhindar had been killed by a snake-bite on their wedding night on orders of Manasā, the goddess of snakes. The angry goddess caused this death to punish Lakhindar's father, Chāndo, for refusing to worship her. The goddess had previously killed Lakhindar's six elder brothers and brought financial ruin on Chāndo by destroying his fourteen merchant ships. After Lakhindar's death, during the long river journey with her husband's decaying body on her lap, Behulā stops at many river ports (*ghaaTs*), and her encounters at these places are described. Behulā eventually is able to find a way to appeal to the angry goddess at a court of the gods, who succeed in persuading the

*The source of this story is a medieval Bengali poem of Ketakā-dāsa, the *Manasā-maṅgal*. According to Edward Dimock (1963), there are many versions of this legend coming from different writers in Bengal; and it may have existed as an oral tradition before it was written down. P.K. Maity also writes about this story and about the cult of Manasa generally (Maity 1966). This goddess "is worshipped for the removal of barrenness, for causing rain, for curing diseases, for wealth… etc." (Maity 1971:29)

**According to Maity, it was common practice (in some unspecified place and time) to place the corpse of someone killed by snake bite on a raft of banana stumps and let it float on a river. This was done in the hope that the body would be found by a person with magical powers (called *ojha*) and brought back to life. (Maity 1971:32)

goddess to revive Lakhindar. Behulā is Manasā's devotee, and she persuades her father-in-law to end his feud with the goddess. (Dimock 1963:195-294)

This story is still well known in the countryside, supporting a sense that the Bengali landscape once was traversed by this brave and virtuous woman. In the original written version, Behulā travels on the Damodār River, which is in West Bengal; but other rivers—and even other types of water bodies—get involved in other versions. For example, in Kolirani Subdistrict of Chandpur District of Bangladesh Behulā's spice-grinding stone is said to be located next to a large water body (*dighi*).* In this area, like other parts of Bangladesh, according to Professor H.K.S. Arefeen, there are other landmarks that also connect Bangladesh's Chandpur District with the "Behulā-Lakhindar" story.

Some South Asian pilgrimage sites are located at waterfalls. Disanayaka explains that in Sri Lanka, at least, the myths and legends associated with waterfalls follow "a conventional motif." The typical story, according to this source, is: "A king goes to war, leaving behind his queen, with the instruction that he would inform his victory by getting his messengers to wave a white flag. By sheer negligence, a black flag is waved and the queen commits suicide by jumping into the fall, as a token of her faith to her husband." (Disanayaka 2000:20) This story surely is associated with more than one waterfall in Sri Lanka, but the inventive devices of South Asian water culture no doubt associate many other stories with waterfalls elsewhere.

Seafaring traders and ocean fisherfolk have folk traditions relating to the ocean, but there are only a few reports describing them. Santa Rama Rau describes beliefs among some Kerala fishermen in a "stern goddess Katalamma." This goddess is seen

*The location is Koilrani Village, Chandini, near Kochua. We thank Professor H.K.S. Arefeen, of the Dhaka University Anthropology Department, for this information, which he shared in a March 2008 conversation.

as a punishing force, the personification of dangers faced daily by fishermen. "She punishes the wicked, dragging them down to her terrible realm and sending sea monsters and serpents to the beach as a warning of her wrath." Katalamma is an enforcer of morality. "The men at sea must be brave and moral. The women on shore must be pure and chaste to guarantee the safety of their men on Katalamma's risky waters." (Rau 1962)

P.K. Maity describes a special rite, called *Bhāduli brata*, performed by Bengali women to ensure their men's safety while at sea. Once "very popular throughout undivided Bengal," this rite involves a complicated series of actions and offerings. The women "dig a miniature excavation to symbolize the sea," create a small mound on which a child-goddess, Bhāduli, is seated. They then clean the area and make rice-flour paste drawings (*alpana*) of small seas around the hole, "so that the total number of seas should be seven."

Then, Maity continues, the women make a drawing of a large river with the same rice-flour paste. "Thirteen mouths of the river should be joined to the aforementioned seas." Other drawings are made of various items, including animals, trees, a hanging bird's nest, and a raft. The women then collect water from a nearby river or tank and bring it to the ritual site in a vessel. Chanting special verses, the women pour this tank or river water on their excavated and painted seas and rivers.

> A miniature boat made of banana bark is set afloat. The goddess Bhāduli is invoked for the safe return of the relatives of the *bratinī* [female ritual actors] through the recitation of the *chhaṛās* [verses]. (Maity 1988:97-99)

Bernadette Gomes's Goa study describes a folk analogy between the sea and the flow of fluids in the human body. She discusses a Goan "folk gynaecology" belief that sea baths are helpful to maintaining the flow of blood in women's bodies. An account given by elderly village women declares that,

> The right days [for sea water baths] are when the water turns a muddy red along the coast. The waters themselves are called as *rogtachem udok* (blood water). They are especially sought as a remedy for painful joints. (Gomes 2005:262-263)

Ponds, Tanks, or Lakes

Throughout South Asia, human-made ponds or tanks are set up near Hindu temples and mosques. These tanks are used for ritual purification before worship and for other purposes. Before the widespread use of tube wells, they often served as reservoirs providing drinking water. They may be inhabited by animals, such as turtles, fish or crocodiles, which are considered to be holy in some way.

Bengali lore endows some types of water bodies, especially large ponds or lakes (Bengali *dighi*), with magical qualities. Most of these were created by revenue-collecting *zamindars* or other types of feudal lords.* Some were dug by saintly Muslim figures, such as Khan Jahan Ali in Bagerhat District or Hazrat Shahjalal in Sylhet, who controlled large tracts of land. The purposes of these water bodies were to provide drinking water and to spread the aura of Muslim saints (*pirs*).**

---

*According to Kränzlin, the right to dig such a pond "remained an exclusive right of the *zamindar* until independence [from Britain] in 1947.... The maintenance of these ponds was the responsibility of the *zamindars*." They were dug and used principally to provide a supply of drinking water. (Kränzlin 2000:74)

**Khan Jahan Ali died in 1459. Khan Jahan Ali's tank or pond is a lake-like tank in front of Khan Jahan Ali's tomb complex in Bagerhat District, Bangladesh. The tank has several marsh crocodiles in it. People believe that if they appease the hunger of these crocodiles with chicken or goats, they will have their hearts'

They still are used by local people for multiple domestic purposes such as bathing, utensil and clothes washing, drinking water, and so on. Throughout the Bengali speaking region and elsewhere in the subcontinent there are a great many myths and legends about such water bodies—their origins, their special powers, and some of their animal inhabitants. Some stories are discussed below, in Chapter 3.

In her Bangladesh study of ponds, Irene Kränzlin (2000:76-77) sums up some widespread ideas about them:

1. Ponds are locations for hidden treasure, which is guarded by a snake.
2. Ponds often require a sacrifice. The pond invokes a respectful reciprocity of man towards nature: to receive good water, the source of life, man has to contribute an equal share. In stories, this often happens through human sacrifices.
3. The pond is an autonomous body, an animated place that links the surface with the underworld and human beings with spirits. The ponds are often inhabited by a crocodile or big turtles, which are transformed spirits that remind people to respect the water. The pond can offer objects and the water can change color or boil.

"The most attractive ponds that are frequently visited in Bangladesh today are those inhabited by big turtles or crocodiles, for example the Bajasid Bustani [Bayejid Bostami] *dighi* in Chittagong or the Khan Jahan Ali *dighi* in Bagherhat (Khulna)," according to Kränzlin. "These ponds go back to historically and religiously important persons, for example to *pirs* (local saints) whose life stories are repeatedly told.... In Karnataka there is a pond where today around 40 devotees daily present

---

desires fulfilled. Whenever anybody wishes to make an offering, the caretaker of the tomb complex, or *mazar*, calls out the crocodiles, shouting "Kalapahar, Dhalapahar, come!" Within a few minutes, the crocodiles make their appearance and swallow the offerings.

Photo 2-4. Feeding large turtles at the tank near the Tomb of Sultan Bayejid Bostami in Chittagong District. Animals in such tanks are regarded as reincarnated spirits capable of granting wishes. (Photo credit: Shireen Akhter)

offerings to Babya, the temple crocodile, in order to receive its blessings."(Kränzlin 2000:76-77)

Animals with spiritual powers are commonly associated with water bodies. For example, Bayejid Bostami, a famous lake in Chittagong District, is inhabited by huge tortoises said to be reincarnated spirits (*jinn*) being punished for mistakes in their previous lives. There is a big tree near the lake. People feed the tortoises, express their wishes, and wash their hands. (Photo 2-4) The tortoises are given much respect. When they die, they are buried in a nearby cemetery (*maazhaar*). They are said to be members of the "Jinn Patrilineage" (*jinn bongsho*).*

Ponds and other water bodies are mentioned frequently in ethnographic studies of Bengali women's lives and rituals. Ellickson (1972), for example, describes repeated trips to ponds to immerse items used in wedding rites.

*Shireen Akhter report, 2007

Blanchet (1984:90) reports a belief that a placenta is best buried near a river or pond to ensure that a new mother's milk supply will be adequate. Rivers, like jungles, she says, are considered "natural," wild or untamed places in comparison to land, which is perceived as "cultural," or under human influence.

Summary

The Bengali-speaking population is part of the South Asian subcontinent. South Asian civilization—a blend of Hindu, Buddhist, Muslim, and other traditions—has a largeand complicated body of knowledge and mythology relating to water. This chapter briefly reviews the cultural history of the Bengali-speaking region and its unique ways of com bining diverse traditions. Early texts identify water as one of four or five cosmic elements. Some ancient humoral medical thinking based on this theory was incorporated into indigenous theories. Humoral medicine still is the basis of folk healing and popular ideas about the human body. The chapter briefly reviews the role of water in South Asian mythology and ways that water-sharing defines social relationships. A short literature review summarizes ethnographic and folkloric information on rivers, rain and ponds/tanks.

3. Bengali Water Lore

There is an abundance of folklore associated with water in the areas we have visited. Stories, sayings and songs use water in symbolic and poetic ways. For example, one day at a riverside gathering in Bera a devotee of a Muslim saint, Pagol Mainuddin, sang a song for us in which the ocean's unexplored depths served as a metaphor for spiritual mysteries. The opening lines were, "You think the ocean is very deep, and there are valuable things in there. It has strong currents and tides. Its depths are not accessible to most of us. You must make yourself into a skilled diver with spiritual powers, if you are to plumb the ocean's depths."* This aspect of water life is important to a great many of the people we have interviewed. Some stories are genuine myths, in the sense that they are believed to be true. In the course of doing fieldwork we have collected at least 18 stories about miracles or unusual water events that were interpreted as signs of the presence of demanding spirits, messages from God, or the forceful personality of water itself.

Pond Myths

Numerous spiritual and social themes can be found in water lore, especially the lore associated with very large ponds (*dighi*) created by zamindars and other feudal lords during the colonial period

*S. Hanchett and Anwar Islam notes, July 2009, Ruppur Union, Bera Subdistrict, Pabna District

and before. One prominent theme in many stories about such ponds is sacrifice. The life-giving water filling a pond requires something to be given in exchange. Three examples follow.

Fuldan Pond, Noakhali District

Once there was a man named Fuldan. He dug a pond, but no water appeared. Then he himself went into the hole he had dug and disappeared. The hole immediately filled up with groundwater. People believe that this was a miracle of God. Ever since then the water of that pond is considered holy. Many people, both male and female, visit the place to seek fulfillment of their wishes. Women from different areas come to visit in hopes of conceiving a child, having a son, finding a missing son, retrieving lost items, such as missing gold ornaments, and curing illness. They bathe in the pond, drink the water, and bring some water to their houses for later use as a healing agent. One woman said she went there after she had a dream in which she was advised to go. Many agree that their wishes have come true after using that holy pond water.[*]

Kamala Dighi, Noakhali District

Around 150 years ago, a zamindar wanted a big pond on his estate, so he told his workers to start digging. But they somehow were unable to start the work. One day he dreamed that the decorative end of his wife's sari had to be buried in a hole, so he ordered some people to do this, and the digging work began. After the pond was dug, however, water still would not come. The zamindar arranged all sorts of rituals and prayer meetings, but nothing seemed to work. Then he had another dream, that if his wife, Kamala, would be the first one to try to take water, it would flow. Many people warned him that it might be dangerous for her to do this, but he ignored their

*Shireen Akhter report, 2007

*advice. Anyway, his wife had to obey her husband, so she went to the pond. As she walked toward the pond, water came up into it. She filled her water pot, but she could not come back. She went under the water and never was found.**

A Delduar Dighi, Tangail District

In Delduar Village there is one dighi *that was dug by a zamindar. His daughter was very ill. All treatment efforts failed to cure her. One night the zamindar was told in a dream that he should dig a big pond and offer it for the use of all the people, not to use it privately. If he did this, his daughter would be cured soon. He did create the pond for the use of all the people, and his daughter recovered.***

Photo 3-1. Sadi Matbar Dighi, Rajoir District (Photo credit: Suzanne Hanchett)

*1997 field notes

**Anwar Islam report, 2007

Five of our 18 stories describe ponds that formerly produced cooking vessels and other utensils when these were needed for large feasts.* While attesting to the spiritual powers of water bodies, these also are moral tales that exhort listeners to be pious, generous, and honest.

Photo 3-2. Bhanga Dighi, Laksham Subdistrict, Comilla District (Photo credit: Suzanne Hanchett)

---

*Four of these are from Comilla District, and one is from Noakhali, but the theme is surely more widespread. The utensil-providing-lake motif was mentioned more than 100 years ago in a report by William Crooke on lore associated with Taroba (or Tadala) Lake, in Chánda District, Central Provinces: "Formerly at the call of pilgrims all necessary vessels rose from the lake, and after being used were washed and returned to the waters. But an evil-minded man at last took those he had received to his house: they quickly vanished, and from that day the mystic provision wholly ceased." (Crooke 1894:32)

Dewan Shahib Pond, in Laksham Subdistrict, Comilla District

This pond (pukur) was very big and surrounded by a wall. The water was white and transparent. If pious people, people with pure hearts, went before the pond and prayed to get utensils for a festival meal, they would find the needed utensils after some time sitting on the bank of the pond. If someone was not pure (paak), the pond would not hear their prayer. Once some people took the utensils rather than returning them to the pond, and the pond stopped responding to anyone's prayers.[*]

Pond at Old Mosque, Dhakshin Narapati Village, Laksham Subdistrict, Comilla District

Elderly people say that they heard from their elders, four generations back, that at one time people got utensils from this pond whenever they needed them for marriages or any other feasts. But one day after a feast, someone stole some utensils, and the pond stopped providing the things anymore. This was a punishment for their misconduct.[**]

Kanchanpur Dorga Bari Dighi

One Noakhali District (Ramganj Subdistrict) pond (dighi) is located at Kanchaanpur Dorga BaaRi. To this place, an especially spiritual Muslim man and his sister came from Baghdad to spread Islam. They died peacefully in this place. The pond, which is located close to the graveyard, used to give forth all kinds of cooking and eating utensils—plates, spoons, and cooking vessels—made of gold

*Tofazzel Hossain Monju report from a July 2009 focus group discussion in Purbo Laksham Union

**Anwar Islam report, July 2009

*and silver every year for use in a festival named Orosh. The items appeared in a boat floating on the water. After using them, people returned them to the boat, which then disappeared beneath the surface. It is said that once a golden spoon was stolen, and the pond never provided the vessels and utensils again. Many people come to the graveyard to have special wishes granted. The water is not drunk; rather, people use it to heal illnesses.**

The motif of providing cooking or eating utensils for large numbers of guests is linked with certain basic values, especially purity and generosity. It also relates to the cultural principle that one gains honor and spiritual merit by giving food to many other people when hosting a feast.** Conversely, each pond (viewed as a spiritual being) punished theft, a violation of the trust implied by the gift of the utensils. This violation spoiled the community's relationship to the pond, in much the same way that theft from a generous host would spoil a human bond.

A large water body, called Ram Shagor, in Dinajpur District, was said to appear miraculously. No one dug it. There never had been a pond there before. It is very large, so its name translates as "Ram's Sea." It is considered to be a "God-gifted pond." A special feature of this water body is that if one person challenges another to walk around it, it will be impossible for that person to complete the circuit. If not challenged in this way, however, one could easily walk around it.***

Other pond myths are as varied as oral traditions can possibly be. Water appears miraculously. Wishes of all sorts are fulfilled. Infertility

---

*Shireen Akhter report, 2007. A *dargā* is a graveyard where Muslim saints (*pirs*) are buried.

**Hanchett 1975 discusses the importance of good hospitality and feasting others to maintenance of a family's social status.

***Information from our colleague, Helen Rahman (1997), whose home district is Rangpur.

is cured. Diseases are healed. Water boils spontaneously or displays other strange qualities. Reincarnated spirits with wish-granting powers live in water. The amazing stories go on and on, entertaining us, spicing up the life of each region, and reminding us of core social values. They also remind us continually that water has a special, life-giving quality that should not be taken for granted.

School Pond

In Bera Subdistrict three men told us a long story about a local pond that originally had been dug by a zamindar, who left the area in 1947. This water body was once called Zamindar Pukur, but it is now called School Pond.*

When the pond was first dug, a live fish called "Gojar" was needed as an offering (utsorgo) *to the pond. So one such fish was decorated with red powder and thrown into the pond. It grew very large. Sometimes it bit people. No one could ever catch that fish. Every year the zamindar made offerings to the fish in the Hindu ceremony called* pujaa. *The name of that* pujaa *was Path Thaakurer Pujaa, and the fish could be seen coming up to take the offerings. People saw that large fish moving around in the water. A wooden image also received offerings as part of this ceremony, which was done in the Bengali month of Bhadro. The deity is called Path Thaakur, which is another name of the Hindu deity, Bhagobun. Path Thaakur is a "living god"* (gagrato debota). *It is made of wood, decorated with colorful designs, square in shape; and its eyes are made of gold and silver. Hindus believe that if someone bathes the image with milk, it becomes alive* (jugroto). *They also believe that the deity will grant any wish. But it is important that one avoid eating fish, meat, or eggs before conducting* pujaa *for this deity. One must only eat vegetables before doing the offering.*

---

*This water body is located in Bhavanipur Village, Bera Subdistrict, Pabna District.

One day, the men told us, someone disrespected the Thaakur by eating fish and egg and then touching the wooden image. "What can a mere piece of wood do?" he asked mockingly. The man became seriously ill. This Thaakur has the ability to enter into a human's body. It entered his body and spoke [through him], saying that he did a very bad thing by not respecting the Thaakur. The boy's mother pleaded with the deity: "Please forgive my son. I will bathe you with milk." So he forgave the boy. The zamindar did offerings to the fish and Thaakur every year, to ensure good fortune (mongaaler jonno). *When the zamindar left, they tried to find someone to do the* pujaa *but no one took responsibility, because the Thaakur image was considered to be a dangerous, living thing. They threw the Thaakur into the pond. No one ever saw it again, nor did they ever see the large fish. Some continued trying to catch the fish, but they could not. The villagers tried to re-excavate the pond, but they just could not do the work. They felt unwell when they tried to dig. They were so uncomfortable, that they stopped digging. They felt that some spiritual force was present. It is said that when the pond was dug, a pipe was installed under the pond, connecting it to the Atrai River. Some say that the fish comes and goes through this channel. "No one can catch the fish because it has supernatural power," the men said.**

Ponds created by Muslim saints (*pirs*) came to be regarded as spiritually powerful because of their association with the saints. One example is Kanchanpur Dorga Bari Dighi, mentioned earlier. Another is Hujurer Pukur (Pir Saheber Pukur) in Patuakhali.

Hujurer Pukur, or Pir Saheber Pukur

In 1997 an imam of the mosque associated with this water body and four other men told us about the history of the saint and the magical properties of the pond. The Jainpur Pir dynasty is well

*Report by Shireen Akhter, 2009

known throughout India and Bangladesh, they explained. After his predecessors had established stations in Rangpur, Dhaka and Bhola, Pir Hafez Maulana Hossain Ahmed took over and shifted to Patuakhali in 1926. He does not stay in Patuakhali permanently. He lives in Jainkati, in India, and comes here in the Bengali months of Falgun and Chaitra, from mid-February to mid-April. When he first shifted his station to Patuakhali, his disciples donated land in Jainkati Town, and a mosque (Jame Mosque) was built on that land with disciples' donations. At the entrance to the mosque compound the pond was formed in a hole dug during construction of the mosque and its plinth. The area of the pond's size is 60 by 40 feet, and it is 10 feet deep. The pond was being used for bathing and for oju, purification before worship. At first, the Pir's disciples came to him to get his blessings. They also brought bottles of water, on which he would blow: that is, "give his *phu*," or push air from his mouth onto the water. This blown-on water was thought to have the power to cure diseases.

One day, while all the people were taking phu *from the Pir in this way, one of the Pir's followers said that the blowing did not actually touch the water in his bottle. So he threw the water from his bottle into the nearby pond. Suddenly the area of the pond where the water had entered started to boil. When the Pir Saheb learned of this, he had the man fill up his bottle from the pond, and the boiling stopped. Ever since then, people started taking water from that pond on the assumption that it could cure diseases. The Pir Saheb asked his disciples to use the water for that purpose, since he does not stay in Patuakhali all the time. The water of the pond is muddy, because rain water falls in it; but there is no connection with the river or tidal flows.**

---

*Report by Nurul Absar, September 1997, Jainkati Town, Patuakhali Pourashava, Ward No. 1

Stories About Rivers

As discussed in Chapter 2, rivers also may be perceived as having special powers as beings in themselves or as the homes of spirits. For example, there is a Hindu pilgrimage site, Langalbandh, on the Brahmaputra River in Narayanganj District. This river is regarded as male, the son (*putra*) of Lord Brahma, who is the Hindu god of creation. Every year on the eighth day of the Bengali month of Chaitra (March-April) people come to dip in the Brahmaputra's waters. This one act is thought to allow pilgrims to be absolved of all sins, and also to acquire multiplied merit equivalent to several pilgrimages.

At another pilgrimage site, Madhob Kunda, in Moulvibazar District, Hindu pilgrims take a special bath—called Baroni Snan—on the 13th day of the Bengali month of Chaitra in a waterfall that flows out of a natural water reservoir (*kunDo*). During this bath, it is believed, all sins are released into the water.*

In Bera Subdistrict we heard the following two different versions of an incident interpreted as evidence that spirits live in the Atrai River.

The Fallen-down Buffalo (*Moish Gaari*)

Version No. 1. *Sixty or 70 years ago during the rainy season an ox cart was trying to cross over the Atrai River on a small mud path. Near the path was a deep pool in the middle of the river. The two bullocks pulling the cart both fell into the hole, pulling the cart in after them. People searched for them but found neither animals nor ox cart. It is assumed that some supernatural beings (*deo*, meaning "deity") live in this deep place. They are not considered dangerous to people, but they did eat those bullocks. Some people think that there is a Muslim priest, Hajrat Khijir,*

*Kazi Rozana Akhter report, 2007

*bullocks for some reason.**

Version No. 2. The second version we heard of this story had the cart crossing the river on a bridge 100 or 200 years ago. *The cart was pulled by a water buffalo, which went off the bridge and was submerged in the river bed. The strange thing is that the river bed was dry at the time, so it should not have been possible for anything to sink down into it. The disappearance of the water buffalo and cart is regarded as evidence that a spirit lives in the river bed. The buffalo died in the fall and was left there, already deeply buried. Perhaps there was a deep water hole (*ghanaa*) at that place. People have been making ritual offerings* (boug) *there ever since.***

Development and Myth

One water event observed in September-October 1988 in a Tangail District wetlands area was interpreted as a sign that God was angry about corruption and other sins.

A village in Tangail District had a road going through its center. It was a marshy area. A 25-meter long concrete bridge had been paid for by foreign aid donors; it allowed huge amounts of water to drain out from marshy areas surrounding the village. One morning during the rainy season, normally a time of flooding, as residents awoke to the twittering of birds, they were surprised to see a three-acre pool, a dou, *under the bridge (A* dou *is a naturally created water body having deep, eerily blue water. It is considered frightening, somewhat ghostly. Residents do not allow their children to go near a* dou, *where they might be attacked by ghosts.). People said it was a 'revenge flood' that created the* dou. *Immediately after the* dou *appeared, people started saying that*

*This appears to be a reference to the same water spirit, Kwaz Khijir, mentioned earlier in Chapter 2.

**Two reports by (1) Shireen Akhter and (2) Suzanne Hanchett and Anwar Islam, 2009, Ruppur Union

*God was punishing the village people for their sins—not praying regularly, taking bribes, and money laundering. Many laid out fruits or lit candles under trees, setting their offerings afloat on banana trunk rafts, hoping to regain God's favor.** (Excerpt from Islam 2012:200)

Folk Sayings

In the course of our interviews and discussions, people often summed up their thoughts and feelings using popular sayings. Some of these concern water directly, and some use water metaphorically. Several people told us that, "Water's other name is life." In one folk saying, the flow of a big river is compared to the worthy character of an unattractive (dark-skinned) girl from a good family line: "A black daughter from a good family is good, and muddy water from a good [meaning: big] river is good" (*jaater meye khaal a bhaalo nadir jal galau o bhaalo).*

We collected some folk sayings from men and women in Bera, both Hindus and Muslims. Many of these sayings instruct women to behave properly and express common problems, fears, and beliefs. Examples include: "Don't show your appearance to a man. This is sin. You will not go to Heaven." "Any unpleasant situation—some damage, an unexpected death, any bad luck—is blamed on a newly married wife." "A pregnant mother should not go out in mid-day or evening, because an invisible spirit or ghost (*petni/pichaash*) could attack both mother and child." "A female can get sick if the end of her sari or scarf (*ornaa*) touches the ground. A ghost or demon could climb up the garment onto her body." "Give the fruits of a new tree to elderly people to eat. If you don't do this, next year's production will be reduced."

---

*Anwar Islam report

Sayings That Relate to Water

"A clay stove (*chulaa*) should be polished with muddy water." (Meaning: Before cooking polish up the clay stove. If it is not done, family members may get sick.)

"Don't urinate into water. It causes pollution, and is a sinful act. *Maa gaŋgaa* (Mother Gangā, a deity/*debotaa*) and Khoaz Khizir are in the water and may curse you."

"Don't stroke your husband with your wet hair. It will reduce his lifetime."

"After bathing, don't let your wet clothes drip water onto your leg. A great man drinks this water."

"Don't clean up cow urine when your body or clothes are wet. It will harm the cow's health." (Hindu saying)

"A dead man drinks water from the burial ground." (Meaning: the dead are revived at sunset—*bhaaraa saandhaaoy*—and come out to drink and eat. So you should not eat or drink at this time of day.)

The sayings about women's wet hair or clothing may refer to concerns about some kind of pollution harming others. That is, water dripping off of a woman's hair, body, or clothes may carry pollution associated with her menstruation or birthing processes.

Summoning Rain

In three of our study districts—Comilla, Pabna, and Tangail—we heard reports of rain-making songs and rituals. In Comilla District, some women said that they perform a special ceremony when they become desperate for the rains to start. The ceremony has symbolic features in common with others described in the folklore literature (Chapter 2). It is called "Wanting Rain" (*meegh maagaa*) and is led by an old woman who has given birth to many children. Her group goes to every door in the village while singing a special song. Each house gives her and her group some rice and seven spoons of water. The water and rice are collected in a big pot and cooked into a rice-lentil preparation, *kichuRi*, which is customarily served during the rainy season. After cooking they distribute the food to all children of the village. Finally, they offer a special prayer to God for rain.* The type of song they sing is called *chhikalii.* The words of one song translate as follows:

Oh Cloud, Our Cloud
Oh Mother of Cloud
Drop rain continuously
On the winnowing-fan made of cane.
Oh Allah Cloud,
Mother of Cloud died.
Where will we find her?
The seven brother fishermen
Going home to make their nets.
Water in drainage channel of aarum *[taro]*
Come Cloud Water
Drive away the brother to that way,
To that way.
Sleepy, sleepy tiger cub,
*Give him rice and send him away.***

In another village in the same region, a similar ceremony was described by a local woman who worked for an NGO as a community health promoter. "There are water scarcities in our village sometimes," she said. "When it is late in the dry season and the expected rain does not come, one elderly woman collects children

*Kazi Rozana Akhter report, 2009. This description is similar in some ways to customs described by folklorists and P.K. Maity, discussed earlier in Chapter 2.

**Bengali lyrics are in Appendix 3.

from each house to sing special folk songs (*jaarii gaan*) to get rain from God. We decorate a girl with flowers, and she carries a flower-covered winnowing fan (*chaalaa)* on her head as she moves around accompanied by many children. The girl sings, 'Allah give us some rain' (*aallaa megh deoy paani deoyree*) and begs for rice, pulses, turmeric, onion, and salt. This is cooked and offered to children as a feast. We believe that this ceremony can bring rain from God. In some years rain has come down very suddenly. We believed this happened because of our rain ceremony."

Another rain-making ceremony was described by a group of women in Bera.* "We sing these songs when there is a drought and our crops are burning up. Seven or eight women and some children gather in a courtyard and pour water on the ground to create mud. Then the women and children—women of all ages, old and young—roll around in the mud singing these songs as they roll. We did this ceremony already three times this year, all during the month of Boishakh [April-May]. We have full confidence that God hears this prayer and responds instantly by sending rain. Rokeya [a spiritual woman with matted hair] is the organizer of the event." The English translation of one song follows:

> *O Sky* (deowaare), *give us some rain.*
> *Pour some continuous rain.*
> *We put some palm leaves on our roof,*
> *But we want so much rain that it will come inside the house.*
> *The* chinaa *[corn] field has a little stream of water* (chinchinaa paani),
> *In the paddy field, water comes up to the knees.*
> *Cutting the drain (laali), and water will flow out to the field.*
> *Who is the* matabor *[leader] of this village?*
> *Hari Krishna*, meghere *[cloud], pour some water.*
> *We have no water for washing rice,*

*Suzanne Hanchett and Anwar Islam notes, 2009

We have no water for washing pulses/legumes.
Crazy Chan (paagol chaan), *you* (tui) *please come to us.*
*I lost all my caste (*jaati*), all my dignity, everything, for you.*
I have offered much to you.
Please come to us, and give us water.

Two rain-making ceremonies described from Tangail District have not been performed in recent years, as far as we know, but one of them is similar to the one described in Bera:

(1) Now at the dry season people feel worried and anxious. They pray for rain. A spiritual guide (*montro data*) performs *vedik montro* (chanted Vedic prayers). A water pot (*kolshi*) is set in a certain place, and holy words (*montro*) are recited. The montro gives the water some "heavenly power." This water is poured all over the earth of a courtyard. The spiritual guide brings young, naked children into the courtyard. He requests the children to roll in the soil and make it muddy: *kaadaai maakha-maakhi*, which means "rubbing mud on the body." This is done in a special place in a village neighborhood (*paaRaa*). The children roll around in the mud. Then the spiritual guide gets rice from everyone in the neighborhood as payment for his service. This is a kind of business for him. Muslims and non-Muslims (Buddhists, Christians, or Hindus) all participate. God is expected to give rain after this ceremony.
(2) *BrishTir jonno doaa* (Prayer for rain). Muslims do a prayer with the same goal. This is just done for Muslims, however. An imam (leader of the mosque) attributes the drought as punishment from God. So he proposes that we go to a field and pray to God for rain water. The imam in the field is accompanied only by men. Women are not allowed. The imam recites some Koranic verses. During

> this prayer, the men all stand in a long queue. Their hands are held palms down in a V-shape rather than in the usual palms-up manner of prayer. In the past, they say a very holy and righteous imam has succeeded in getting God to provide rain instantly by performing this ritual.

In three of the four rain-making ceremonies described above, one clear theme is fertility. A woman with many children leads the ceremony in one Comilla description. Children themselves (emblems of human fertility) are essential to the ceremony in all three. Rain water falling on the earth of agricultural fields is the desired goal, and this surely explains why two of the ceremonies include mud—a mixture of earth and water. These rituals are similar in these and other details (including nudity) to some of those described in Shankar Sen Gupta's book, *Rain in Indian Life and Lore*, and mentioned in Chapter 2.

Summary

This chapter discusses some myths or stories and sayings associated with water and water bodies (ponds, tanks, lakes). These can be understood at several levels, but most serve to reproduce social values. Some themes are sacrifice, exchange, and honor. Out of 18 myths told to us, five were about very large ponds providing vessels needed for feasting. Common sayings are, "Water is life" or "water's other name is life." There are songs and rituals —not frequently performed nowadays—to bring rain.

4. Perceptions of Water

*"We judge water on its taste. Our tongue searches for taste. And we should consider its effects on our health."**

The supply of domestic water affects family health and family life. People's judgments about water—which types are good or bad for which purposes—strongly influence how domestic water is used. The effectiveness of domestic water services and programs, therefore, depends to a large extent on whether such judgments are taken into consideration by planners and service providers. This chapter describes the language and concepts that organize Bengali villagers' thinking about—and uses of—domestic water sources.

Water Language

The words used to describe water reveal much about how it is understood. People's habits, tastes and beliefs are expressed partly through the kinds of things they name and how they talk about their experiences. There are differences in the water language of different regions, ethnic groups, and religious communities. For example, when speaking of "water" in general, Bangladesh Hindus tend to use the ancient word *jal*, which has religious overtones, and Muslims mostly use the word *paani*. The

*Tofazzel Hossain Monju notes, Laksham, 2009

word for water changes to *haani* in dialects of Noakhali, Feni and Laksmipur Districts, and in parts of Comilla.

Asking "What are all the kinds of water?" during 2009, we collected some detailed word lists with approximately 10 focus groups and in several individual interviews. These lists were supplemented by our own knowledge as people who grew up in some of the areas and by a few interviews in other regions. In arsenic-affected areas, we have observed some new water vocabulary developing over the past 10 or 20 years. For example, people now describe their tube well water as either "arsenic free" (*aarshonik-mukto*) or "arsenic contaminated" (*aarshonik-jukto*). The most complete information on water language that we have was collected in Bera Subdistrict. Our full water vocabulary list is in Appendix 2.

Types of Surface Water

Numerous types of water body are named. The distinctions reflect the fact that differently named water types have different uses. Large canals (*khaal*) wind through many villages. They are used to bathe cattle, to culture fish, and sometimes for domestic purposes. Rivers (*nadii* or *gaaŋ*) are never far away. Extensive and lush wetlands (*bil* or *baor*) support fisheries and large populations of wild birds. "Indeed," as Jashim Uddin and Chowdhury point out, "in the rainy season, about half of the country could be classified as wetland." (1999:28)

Ponds are essential to domestic life in our working area. Linguistic distinctions are made between ponds of different sizes. There is regional variation in the words for "pond" or "tank," but in all the districts where we have worked people distinguish very large water bodies from small or medium sized ones. Very large ponds mostly are used for commercial fish culture, and the old ones often have myths or legends associated with them, as discussed in Chapter 3. Smaller ponds are used mostly for daily

Photo 4-1. A typical canal *(khaal)* crossing a village in the dryseason. Bhanga Subdistrict, Faridpur District (Photo credit: Kazi Rozana Akhter)

household purposes, such as bathing or cleaning utensils, and also for small-scale fish culture. A small pond is generally smaller than one acre and and surrounded by a raised earthen boundary, called *paaR*.

The average size of a fish pond or tank in the Bangladesh floodplain is 0.11 hectares, or approximately 0.7 acres, according to Jashim Uddin and Chowdhury (1999:29).

Photo 4-2. Small water bodies, all called *maiThaal* in Bera Subdistrict, Pabna District (Photo credit: Shireen Akhter)

Photo 4-3. Irrigation channel *(naalaa)* going out into agricultural fields

Photo 4-4. A very small water body in a field is referred to as *gaaRaa* in Comilla District. (Photo credit: Suzanne Hanchett)

Ponds of all sizes are distinguished from very small water bodies (puddles, rivulets, or small pools) found in ditches or fields during the rainy season. These small water bodies are used to water family vegetable gardens, possibly for seasonal fish culture, and sometimes for feminine hygiene purposes, such as washing menstrual cloths.

Water bodies are distinguished by their shapes as well as their sizes. Some people we interviewed made a point of mentioning that some ponds are round in shape, while others have irregular dimensions. The practical or symbolic importance of shape is not clear.

The presence or absence of steps or a platform (*ghaaT*) at the edge of a pond has social significance. We were told that a proper *ghaaT* is a sign of high status. Some say, "We only give our daughters in marriage to families who have a household latrine and a *ghaaT*." Poorer families are not able to maintain this standard, at least where the *ghaaT* is concerned. The best quality

Photo 4-5. Village pond with two concrete *ghaaTs*, constructed by a former zamindar. Bera Subdistrict, 2009 (Photo credit: Shireen Akhter)

Photo 4-6. Pond with two *ghaaTs*, one made of concrete (far right) and the other crudely made with logs (center left) (Photo credit: Shireen Akhter)

Photo 4-7. It is common practice to bathe young children in water that has been warmed in the sun for a while. (Photo credit: Shireen Akhter, Bera 2009)

Photo 4-8. Children bathing and playing in a water body near open ("hanging") latrines. Narayanganj District, 2001 (Photo credit: Shireen Akhter)

ghaaT is made of brick or concrete, which is somewhat expensive. Less valued structures are made of wooden logs*.

The people we interviewed were not in total agreement about water vocabulary. In one all-male Comilla group discussion, participants debated about whether a certain large pond should be called a *dighi* or not. One participant in an all-male group discussion said, "We have no *dighi* in our village." Another disagreed with him, saying, "The pond of the Northern Neighborhood (*uttar paaRa*) is a *dighi*." The first replied, "It is a large pond (*boRo pukur*) but not a *dighi*." Two others agreed with the second man, saying that calling a large pond a *dighi* is an old custom here, but "the idea is still valid." Three others strongly opposed this idea, saying that a *dighi* is a special water body associated with religious faith and a certain type of management. They gave the example of Khan Jhan Ali Dighi, which is a special place, saying that "the water of this *dighi* is serene." "The Northern Neighborhood pond is large," they said, "but it lacks many features of a *dighi*."

Water Quality Distinctions

Rural people are very eager to discuss the qualities of different types of water. A larger volume of water is said to be cleaner and more "pure" than a smaller volume of water. If water is clear or transparent (*farshaa, shosscho,* or *TalTala*),** it is considered to be better than water that is not transparent. Standing water is viewed very differently than flowing water. Flowing water is considered to be cleaner and more desirable, while standing water, especially if it has a relatively small volume, is prone to

* Ports at river banks are also called *ghaaTs*.

*******Farshaa* is best translated as "white," as in "white-colored skin," while *shosscho* means that water is extremely clear, so clear that it reflects back a sharp image when one looks at it. *TalTala* refers to normal water that is fresh and has no extra things in it.

getting "stale" (*baashi*), "dirty" (*aaporishkaar*), or "disgusting" (*noshTo*). One Hindu woman said in a group discussion, "River water (*gaaŋger jal*) is very good. The river takes away all bad things with its waves."*

Heavy or Light Another perceived quality of water is its "heaviness" or "lightness." In Bengali *bhaari, paatla,* or *haalkaa,* respectively, mean "heavy," "light," or "thin." The words can refer literally to the weight or intensity of something: *e.g.*, "heavy rain" is *bhaari-brishTi*. One group said that "heavy" also can mean "thick" (*ghano*), and "light" can also mean "thin."** The two words also are used in a more subtle sense in relation to different types of water:

(1) "If you take tube well water into your mouth, you will feel that it has weight (*bhaari*), but Sidko [neighborhood arsenic removal plant] water is light (*paatlaa*). This filter water is like river water (*gher jal*). When we drink filter water our soul becomes cool (peaceful). We may live without rice for a time but not one minute without water."***
(2) In general pond water is preferred for cooking because it is "thin" (*haalkaa*), that is, good for boiling rice and pulses. It will not change the color or taste of cooked food. Tube well water, on the other hand, is considered to be "heavy."****

Color and Age. In both Comilla and Pabna, people spoke positively of deep water as having a nice, "black" color. "Older,"

*Bera Subdistrict, Shireen Akhter notes, 2009.

**Laksham village group interview

***Laksham Subdistrict village, Tofazzel Hossain Monju group interview, July 2009

****Kazi Rozana Akhter notes, 2007

stagnant, or dirty water was described as red in color. Good water is more often spoken of as being "white" in color and "transparent."

Black is rarely associated with an auspicious or positive description, but in the case of water, it is positive. Two men in Homna Subdistrict explained this to us: "In Ashar and Srabon months [mid-June to August, the monsoon season] there is much water. It has a nice, black color (*kaalo raŋ*). This is 'youthful water' (*jobankaal*), which has that black color because it is so nice and deep. Water in the big rivers (*boRo-boRo nadi*) is very deep. But in Ashin and Kartik months (September to November), when there is less water, the color changes to red. This is water in its 'old age' (*paani briddhokaal/marankaal/sheshkaal*). It has red color and smells bad. It has bugs and dirt in it."*

One group of women in Laksham Subdistrict struggled to sort out their various perceptions and understandings about the best ways to collect and store drinking water. They were finding it difficult to distinguish iron from arsenic. Their comments revealed profound confusion. Iron, which leaves a red-colored residue, is a common problem in their tube well water. Red paint, however, also is used to mark arsenic-affected wells. Their conversation occurred in response to our question, "How do you know which water has arsenic and which has iron?" Here are some of their confused answers: Rongmala said, "One type of oil (*tel*) floats on arsenic water. If you keep it in an aluminum pot, it will be red in color. If you boil this water, the color will change to white (*shaadaa*), and the water will be arsenic-free. But iron water is like cow urine (*garur chanaar moto*)**. Iron is dangerous. It makes all the pots red when we store it. Many types of diseases

---

*Comments of two men in a group discussion, Homna Subdistrict, Comilla District, Anwar Islam notes, June 2009

**The meaning of this comparison (iron water and cow urine) is not clear. It may refer to the common assumption—especially among Hindus -- that cow urine is a powerful substance.

are caused by iron." Rabeya added that, "Arsenic water is white." Rezia Khatun said, "When you pump tube well water, white water comes out at first. If you keep this water in a pot, you can see a certain type of oil floating on the water. If you keep it for a long time, you see the water changes to red." Disagreeing with Rabeya, she added the comment that, "White water is arsenic free."[*]

Gender. Rivers are usually assumed to be female, if their gender is considered. But in Bangladesh there are some male rivers as well. The differences between male and female rivers (*nad* vs. *nadi*) are taught mainly in schools. Some of us were told as children that the waves and currents of a female river are strong but in a good way. Male rivers were said to have more destructive qualities, including very rough waves (*taraŋgaa*).

Some men in Bera told us that the Atrai River, which is considered to be male, came into existence suddenly in one night. How it happened, no one knows. This was more than 100 or 200 years ago. The Jamuna River is considered to be female. There are some songs about that. One is about the Atrai seeking to join with his lover, the Jamuna.[**] Until 40 or 50 years ago, the river was full of water year-round. It had a strong current. Now it has little water for six months of the year. It only has water during the monsoon period, from Ashar to Agrahaayaan (June to December).

Absorptive Capacity. This property of water is implied in a folk healing practice called "reading the water" (*paani paRaa*). This widespread practice consists of speaking holy words over water, which is assumed to absorb their power. This shows that water is perceived as capable of absorbing spiritual or verbal

*Kazi Rozana Akhter field notes, 2009

**Suzanne Hanchett field report, Pabna District, July 2009. The Brahmaputra also is identified as a male river.

things, much as it absorbs or dissolves physical materials. This property of water was mentioned in Chapter 3, in connection with the Hujurer Pukur (Pir Saheber Puker) in Patuakhali, where a Pir made *phu*, or puffed/blew his blessings onto the water.

Cleanliness and Purity. Cleanliness and purity of water are related concepts but not identical. Clean water is considered safe and hygienic for consumption, cooking, the daily bath, or other such purposes. Generally used terms for "clean" are *porishkaar* (literally "clean") and *bhaalo* ("good") while "dirty" is *aaporish-kaar*. Both Hindus and Muslims use these words.

Purity, on the other hand, is a spiritual-*cum*-physical condition, either temporary or permanent, that has social and personal meanings beyond simple hygiene (see Chapter 2). Impurity results from various causes, such as menstruation, childbirth, contact with feces or sexual fluids, and being born into certain castes. It is a dangerous condition resulting in some degree of social isolation and the need to repair or purify if possible.

Two sets of words are used to refer to "purity" and "impurity." Bangladesh Muslims prefer to speak of *paak* and *naapaak*, respectively, while Bangladesh Hindus are more likely to use the words *pabitro and aapabitro.*

Clean water is important for most routine domestic functions. Pure water is needed for ritual ablutions of both Muslims and Hindus, and for Hindu offering rituals (*pujaa*) Cultural notions of "purity" and "pollution" are associated closely with ideas of water's spiritual power, as water is considered to be the ultimate purifying element. One group of Muslim women in Laksham Subdistrict expressed a widespread view, that the purest water in the world is *Jamjam* (also called *Jumjum*) water from Mecca.*

One group of Hindu women in Comilla District said they prefer to use tube well water for *pujaa* offerings rather than pond

*Jamjam water is discussed in Chapter 2.

water because it comes from underground. That is, it is isolated from contact with polluting influences. Their tube well water is contaminated with arsenic, but this is not considered to "pollute" the water they use for ritual purposes.* We heard the same comment from some Muslim men in a different part of the same district. "In our ritual purification procedure (*oju*) we try to use only clean and fresh water," they said. "So we avoid polluted water and use tube well water. For our special Milad prayers, we use tube well water to drink and wash."**

There are customary ways of purifying ponds in the spiritual sense, but it is not clear from this study how often people actually do try to purify their ponds. One Tangail District pond purification method uses the folk healing technique of "reading" holy words into water. In the case of pond purification, the pond itself gets sprinkled with water into which holy words have been "read." The Muslim practice is that a man walks down the *ghaaT* (stairway) toward the pond. He takes some water in his hands. He then recites some Koranic verses over the water and throws it into the pond, to purify the pond. This is done for ponds, rivers, or canals (*khaal*). Women may also do this, following the same procedure.*** A common method of disinfecting pond water is to put lime (calcium carbonate) into the pond. This, however, is not regarded as purification in the spiritual sense.

Concerns about Purity and Cleanliness of Pond Water

In the Comilla District of southeastern Bangladesh, two groups of men discussed with us their need for "pure" water sources to

---

*Muradnagar Subdistrict, mid-2009

**Tofazzel Hossain Monju and Anwar Islam notes, Laksham Subdistrict, July 2009.

***Anwar Islam report, Delduar Subdistrict, Tangail District, 2007

use in ablutions (*oju*) before Islamic prayers. If women bathe in a pond, another men's group agreed, the water is unsuitable for ablutions, because no one knows whether they are in a "pure" (*paak*) or "impure" (*naapaak*) condition. One Laksham Subdistrict men's discussion group agreed, "So we don't do *oju* in a pond where women take a bath."

These comments reflect an intense concern about religiously defined "purity" (*paak* or *pabitro*) and "pollution" (*naapaak* or *aapabitro*), which are sometimes called "ritual purity/pollution" in the scholarly literature New mothers and newborn infants are not allowed to use common ponds for bathing, because their blood is considered extremely "polluting" in this sense.* The blood of menstruating women or new mothers is generally assumed to cause both physical and spiritual damage, if not death, to men. There usually are separate bathing arrangements behind living areas for women in these dangerous states. In some communities with only one pond there may be separate sets of steps down into the pond: one set for use by men and the other by women.

Debris getting into ponds is a common complaint in all of the areas where we have worked. This is another type of pollution, which is similar to the environmentalists' concept. In this case, unlike religiously defined pollution, "clean" (*porishkaar* in Bengali) or "dirty" (*aaporishkaar*) are generally suitable English translations when the problem is dead leaves, paper, or plastic trash.

Human feces, however, once again bring up the ritual pollution concern in areas where there is some open defecation or flooding that causes spread of feces out of pit latrines. A young man in a 2009 Comilla District (Laksham Subdistrict) focus group observed that some ponds are more protected than others. Those with high banks preventing garbage and feces from entering during the rains are "pure" (*paak*), he said. Some ponds are

*These restrictions are observed by both Hindus and Muslims.

not well protected, others concurred, or they have open latrines extending out over them.* Such "impure water" (*naapaak paani*) is considered unusable for ablutions or any kind of religious purpose. "Clean water, religion-wise, is good," he said.

During the same group discussion, another group participant commented that, "Tube well water is not polluted, even if it has arsenic in it. The arsenic does not cause any problem with using it for *oju* and [Muslim] prayers (*naamaaz*)." Such comments strongly suggest that arsenic is regarded more like dirt than like the much-feared agents of ritual pollution, such as menstrual blood.

Although concern about both cleanliness and purity of water bodies seems to be nearly universal, people are not always able to maintain their preferred standards. Urban squatter settlements are especially challenging because of crowding and insecure land tenure. Slum-dwellers therefore often must tolerate less-than-ideal water conditions, although some organizations make significant efforts to help them gain access to safe water and sanitation. Photo 4-8 shows children bathing and playing in water into which latrines drain. We encountered this situation in Narayanganj District in 2001. Although Bangladesh has made significant strides in sanitation coverage since that time, squatter settlements still exist, and such conditions have not disappeared.

Water Qualities Related to Humoral Theories

As discussed in Chapter 2, theories of humoral medicine still survive in popular health beliefs, albeit in modified and flexible forms. Certain prevalent ideas about water relate to qualities described in ancient texts: especially "wetness," "hot," and "cold."

---

*As sanitation awareness has spread in Bangladesh, there are fewer and fewer such latrines to be seen extending out over village ponds or canals but they have not disappeared entirely.

Wetness. Waters and other things are said to differ in their degrees of "wetness," or *bhijaa* in Bengali. For bathing, very "wet" water is best. But "wetness" is associated with "cold" and thus also may cause certain types of illness. Women in 2009 discussions in Comilla and Pabna Districts made the following comments:

(1) "The *bhijaa* water is more cold (*ThaanDaa*) than normal cold water. The water which is more *bhijaa*—it feels very cold. If you put your hand into *bhijaa* water, it makes your whole body shiver. ...Ponds that are surrounded by trees, those where the sun cannot reach to the bottom of the ponds, their water also is *bhijaa*. This water is not good for health. It causes cold (*ThaanDaa*) and, black fever (*kaala-aazaar*)."* (2) "Rain water is more *bhijaa* than other water. It wets the whole body. Tube well water also wets more because it is cold. Pond water is less wet when it becomes heated." (3) "River water never becomes hot. River water is flowing water (*cholti jal*). A flowing river remains cold in all seasons. In the summer, the river water becomes even colder. River water wets the body with peace. If you immerse your body in the river, the body gets fully wet."**

"Wetness" was said to be associated with rot or spoilage in two Comilla District group discussions with both men and women. Comments were these: (1) "Some hybrid rice varieties and parboiled rice become *bhijaa* immediately after they are cooked. These cannot keep for a long time." (2) "Green jackfruit and lentil sauce (*Daal*) made from beans—both of these become *bhijaa* very soon after cooking. Water floats on the curry, and it smells bad. This type of food is very bad for the stomach and causes diarrhea."*** (3) "Food cooked in tube well water soon becomes

---

*Group interview with women in Laksham Subdistrict, Comilla District, facilitated by Kazi Rozana Akhter, July 2009

**Quotes (2) & (3): Group interview with Hindu women in Bera Subdistrict, Pabna District, Shireen Akhter facilitator, July 2009

***Quotes (1) & (2): Group interview with women in Laksham Subdistrict, Co-

rotten. The taste changes. It happens to rice, curry, and cooked vegetables. The color of cooked vegetables changes. Within an hour cooked rice becomes *bhijaa.*" (4) "We always prefer to cook with pond water. The color and taste of food cooked in tube well water is not acceptable to us. Besides, this cooked food quickly becomes "wet," and within a short time it becomes unsuitable for eating.

In the dry season, when there is a severe water scarcity, then people have to use tube well water for cooking. But in the rainy season people use rain water for cooking. They store rain water for cooking more than for drinking, because in the rainy season so many dirty things wash over the bank into a pond, that people want to avoid pond water even for cooking. Rain water also serves as a source of drinking water."*

Hot and Cold. Heat or cold in this sense can refer either to measurable, physical temperature or to presumed "hot/cold" qualities unrelated to temperature. Both, however, are assumed to produce certain physiological and health effects. As George Foster explains, these are entirely different concepts despite the fact that some of the same words are used to communicate about them. Unlike physical temperature, "hot" and "cold" in the humoral sense are best described as what he calls "metaphorical" qualities—properties of material items, properties which never change. Some measurably hot items, such as boiled rice, are believed to be "metaphorically" cold in their physiological effects. Conversely, other foods, such as eggs, milk, and certain types of lentils are perceived to be metaphorically "hot" even when they are thermally cold. (Jellife 1957; Foster 1994) Most in-depth studies agree that variation in belief is normal. That

---

milla District, facilitated by Kazi Rozana Akhter, July 2009

*Quotes (3) & (4): village group interview with men, Laksham Subdistrict, Comilla District, Tofazzel Hossain Monju facilitator (July 2009)

is, within populations subscribing to this system people may disagree about whether a certain food is "hot," "cold" or "neutral," although there are certain foods on which nearly everyone agrees.* So, while detailed ideas can vary, cultural assumptions about "hot" and "cold" and health are pervasive.

As discussed in Chapter 2, a goal of popular health practices and folk healing is to maintain hot-cold balance, as illness often is perceived as resulting from a physiological excess of one or the other. As George Foster points out, a vulnerable person is thought to need protection from sudden or extreme changes in the hot-cold balance in the body, or what he calls hot-cold "insults" assumed to cause illness. One group of men we met in Laksham Subdistrict agreed that excessive "heat" or "cold"—whether in weather, food, or water—is harmful to health.**

Najma Rizvi's study of food and nutrition in Dhaka District explains further: "In Bangladesh, as in other parts of Indian subcontinent, foods are classified into hot and cold types.... Neither the actual temperature, nor any other observable or taste-related factors determine the hot/cold status of a particular food. The "hot" and "cold" properties exist in varying proportions in all life forms—both plant and animal. The people studied seem to view "hot" and "cold" foods in terms of ascending or descending degrees of hot/cold properties....The pattern of hot/cold categorization shows no variation along income groups. Rural/urban differences are also minimal.... Bananas and all citrus fruits are considered "cold" and therefore unsuitable for young children and mothers in the post-partum (cold) state."*** (Rizvi 1979)

---

*Foster 1994

**Tofazzel Hossain Monju and Anwar Islam interview notes, Laksham Subdistrict group discussion, 1 July 2009

***Rizvi goes on to explain that a local fish called *hilsha* (*Tenualosa* ilisha) and shrimp are both considered "hot," while two other types of fish (*boal* [*Wallago attu*] and *puti* [*Puntius spp*].) are "cold," "to be avoided if one is suffering from 'cold' illnesses such as flu and asthma. In the meat group, beef and duck meat are

People in the Bangladesh villages we visited do consider the "hot" or "cold" quality of water,* although water is regarded somewhat differently from metaphorically "hot" or "cold" foods. Water—unlike certain foods—is not generally viewed as having a fixed, unchanging "hot" or "cold" quality or physiological effect. (Certain water sources, however, are regarded in this way by some people.) In the case of water it is the physical temperature that is emphasized. Like a living being, water is influenced by the qualities of its environment.

The humoral theory, however, still is evident in the idea that water temperature has serious health effects, influencing the human body's all-important "hot-cold" balance. Some water sources or water bodies are thought to be so cold, that they must be avoided at times. Two studies have found that boiling water is thought to make it less "wet" and "raw" (*kaãcha*), and thus good for a new mother, whose condition "requires drying and healing foods."**

As will be discussed in Chapter 5, the drinking, bathing, and eating practices of vulnerable people, such as young children or new mothers, are carefully monitored in an attempt to avoid disturbing bodily balance by exposure to too much heat or cold. For example, participants in one village group in Comilla District said, "We avoid 'cold' food. It is 'rotten' (*baashi*). 'Cold' (*ThaaNDaa*) is bad for health." Although opinions can differ as to the presumed heating or cooling qualities of some specific items, anxiety and concern about this matter are nearly universal, especially among parents of young children.

---

considered 'hot' and chicken and mutton are usually 'cool'.... Mango is 'cool', but jackfruit and lichees are 'hot.'" (Rizvi 1979:157-158) For scientific names see http://en.wikipedia.org/wiki/List_of_fishes_in_Bangladesh.

*According to Zeitlyn and Islam's Matlab Subdistrict research (1990), the "hot" and "cold" properties of water are of more concern than the "hot" or "cold" properties of foods.

**Zeitlyn 1993; see also Maloney et al. 1981:131

Participants in one Comilla group discussion expressed these feelings when they explained to us that, "Too much hot or too much cold are both injurious to health. We don't give young children, those aged one to four, any boiled water during illness. We give them water that is moderate ly hot. We keep water in the sun then give it to them. We avoid giving any cold water to children, especially when they are ill. We believe that a child's bodily system is more vulnerable to cold than to hot things. We have practical experience with how cold leads to children having pneumonia, cough, and other serious diseases." Two participants said, "Initially we heard that filtered water [i.e., water treated with an arsenic removal filter] is a little bit colder than tube well water, so mothers should avoid giving filtered water to children. But we don't know for sure."*

Fear of the presumed dangers of cold has come into play during the introduction of other new water supply technologies in Bangladesh rural areas. Zeitlyn and Islam's 1990 study mentioned this in connection with the tube well, a new technology at that time: "...People in the Chandpur area have accepted that the widely provided tube wells give good water for drinking but they do not use it for any other purposes because its taste, colour and temperature are considered inferior." These villagers "never bathe in tube well water because it is perceived to be more cooling than pond water." This source emphasizes that hot and cold refer "not only to temperatures measurable by thermometer, but to the humoral system.**

Between 2006 and 2009, household and community filters were the new technology, introduced to help solve the problem of arsenic in drinking water. Some—but not all—interviewees expressed concern that filter water was too cold. A group of

---

*Tofazzel Hossain Monju notes, Comilla District, 2009

**Zeitlyn and Islam 1990:66-67

women in a Hindu Weaver neighborhood in Bera, for example, expressed the view that, "In winter tube well water is hot, because it comes from underground. In the winter season, no one drinks arsenic-removal filter water directly. Some heat it on the stove before they drink it, if they drink it at all." When young children get sick, mothers offer them tube well water to drink. One woman The women said, "We also drink tube well water when our infant gets a sickness related to cold."* One woman said she normally gives her young child filtered water, but when he gets any kind of cold-related sickness, she gives him tube well water. The women said, "We also drink tube well water when our infant gets a sickness related to cold."**

Unlike these women, who were using household filters, another group of women in the same Bera village felt that the water of the community arsenic-removal filter (Sidko) was more "hot" than tube well water, perhaps too heating at times. "When a child gets cold, he is given Sidko filter water, which we warm up before giving to the child," Hindu community women said. "Since our tube well was screened [for arsenic], we have been trying to avoid drinking tube well water. We made our own *chaari* filters and drank tube well water filtered with that before the Sidko was installed.*** Sidko water is too hot in the summer season, so we keep it in a clay pot (*kolshi*) to cool it. Tube well water is cold in all seasons. Sidko water is more hot than tube well water any way, and we give our children Sidko water."

Most Muslim women in the same village had a different idea about tube well water. All seemed to believe that tube well water was "hot" in winter, and thus good for children. If a child was

---

*Shireen Akhter notes, Bera, 2009. Mothers' and infants' physiological states are presumed to be connected. See Chapter 5.

**Shireen Akhter notes, Bera, 2009. Mothers' and infants' physiological states are presumed to be connected. See Chapter 5.

***Photo 5-3 shows a home-made *chaari* filter.

sick, they never offered filter water because it was considered too cold. Breastfeeding mothers were said to drink tube well water, too, just after birth or if a baby was sick, because it was considered more physiologically heating than filter water.*

A group of women in Comilla District (Laksham Subdistrict) said that they had changed their ideas about tube well water being too cold in the years since tube wells were first introduced. "A new mother drinks only tube well water," they said, "but she cannot drink it directly from the pump. The water has to be heated for some time before the mother can drink it." In earlier times, women drank pond water, not tube well water, after giving birth. They thought that tube well water was colder than pond water. During illnesses such as coughs, cold and fever, children aged two years and older drink tube well water in the summer and warmed-up tube well water in the winter. But infants under one year of age get warmed water in all seasons. Mothers think that if they drink water directly from the tube well, it could harm their health.

Even though these various groups and individuals do not all agree on which water is "hot" or "cold" enough for vulnerable people, they do share a concern about the physiological effects of these presumed qualities.

"Cold" is not always a negative quality. Some adults in the places we visited expressed a strong preference for tube well water that is cold in temperature and thus refreshing for them to drink. The following examples from Laksham (a place with a serious arsenic problem) demonstrate this. Participants in an all-male discussion group said, "We have to consider our available water sources, those which we are entitled to use. Now we generally drink water from various kinds of tube wells. Sometimes we boil pond water, make it cool, and then drink it. Some times

---

*Shireen Akhter notes, Bera, 2009. None of the Muslim women in this group interview had ever been to school, while the Hindu women were mostly educated at least through Class 6 to 8.

it is rain water, and now it's water from [arsenic removal] filters. It is very difficult to select the best drinking water source. Every source has merits and demerits." One man said, "To me, the best drinking water I have had is from a deep tube well. It is so cold and so clean, that my whole body and mind become cool. I get it from Laksham, from my relative's house. The tube well is 1100 feet deep." The other discussion group participants supported him and said, "Yes, there are some very good tube wells, and their water is the best—no iron, nothing is bad." The men agreed, "We like pond water least. Pond water is the worst for drinking even after being boiled and made cool."* Participants in this group said they consider each tube well water's "hot and cold nature" when deciding how to use it. People feel some tube wells' water is hotter than others. So they prefer the tube wells whose water they perceive to be less "hot," especially in the summer.**

Ideas about Rain

Distinctions are made among many types of rainfall: light *vs.* heavy rains, rains lasting a long time, and so on (see Appendix 2). As with surface water types, different kinds of rainfall have different implications for human life. In this region agriculture depends on it, of course. But too much rain or untimely rain can harm crops and damage roads or buildings. Severe floods and cyclones, of course, pose terrible challenges and may inflict considerable damage on rural communities.

Of the three principal sources of domestic supply water (surface water, underground aquifers, and rain), rain is spoken of as

*Tofazzel Hossain notes, group discussion, Laksham Subdistrict

**Tofazzel Hossain Monju notes, Homna Subdistrict, June 2009. While promoting the use of tube wells, public and private agencies conducted intensive campaigns to persuade villagers that pond water was dirty and otherwise disgusting, thus unsuitable for drinking. Comments from this group seem to reflect acceptance of this message.

the purest and most trouble-free. A group of women in Homna Subdistrict agreed that, "Rain water is correct/holy (*shudhyaa*), and that rain water is more pure than tube well water."* Coming from the sky and clouds gives it a quality of holiness. One word for rain, then, is "God's water" (*aallaar paani).* A village group in Homna told us, "When there is no rain, we always worry. Forty or fifty people gather together and pray to Allah. Our prayer makes weather in the sky (*aabuhaawaa).*" Some related comments in Bera were these: "After six months of water scarcity, new rains are appreciated more than gold or diamonds"; "We get rain as if we have found treasure (*kijaani paaichi*)"; and "Rain water is problem-free (*doosh-naai*)."

The people we interviewed did not all agree that rain water is suitable for drinking. In Bera, for example, one group said they do not like to drink it because of its "salty" (*paanshe*) taste but another group in a different section of the village said it tastes "sweet," meaning "good."** A women's group in Homna agreed that they do not like to store rain water for a long time, if they are going to drink it. But they did not specify how long it would be suitable to keep it. They were opposed to the customary practice of keeping drinking water in clay pots. They preferred to use aluminum or steel ones for this purpose.***

In Laksham and Homna Subdistricts, while most people use rain water for cooking during the monsoon season, they have only recently begun to store it for drinking. This may be a result of introduction of rain water harvesting units as a way to avoid drinking arsenic-contaminated tube well water.

---

*Kazi Rozana Akhter notes

**Describing good-tasting water as "sweet" (*mishTi*) is common in several districts of Bangladesh.

***Kazi Rozana Akhter notes, 2009

Photo 4-9. Rain water collection from a corrugated tin roof
(Photo credit: Suzanne Hanchett)

The Power of Words

Language is more than words. Speech is a type of human *action*, which may be thought to be effective in eitherpositive or negative ways. Prayers, spoken or sung, are widely assumed to influence the weather and other cosmic processes. This power is respected and regarded as being somewhat dangerous, as we saw one day in Bera while interviewing a spiritual woman with matted hair, who sang some rain-making songs for us (See Chapter 3). After she had sung for a few minutes, her village neighbors warned her to stop singing, lest the song bring too much rain too early in the year.

The power of speech, especially women's speech, came up again during a group discussion about hail (*shiil*) in a Homna village compound. Hail, they told us, can seriously damage houses and crops. They had a few different Bengali words for hail. One of these was considered dangerous to say, especially if said by women. Our interview went like his: We asked, "Are there any stories *(golpo)* about hail?" A man said, "Hail water destroys our crops. So we're always worried when it omes. We pray for the hail to stop. When it falls on the earth, it makes little pits or holes (*kuus*). It also breaks tin. That's another problem. The local word for hail is *sooraa* or *saairaa*. If women say the word *shiil*, it will come again and again. This word should be avoided, especially not spoken by women. The word *sooraa* is preferred."* He showed us a coconut tree behind the compound that recently had been destroyed by hail. This group also showed us a single pot which they had set at the boundary of the homestead to somehow fend off hail damage (See Photo 4-10).

Comments on hail were initially positive in a women's discussion group in Laksham. They said they use the word *shiil* or *shiiler paani*. They described hail as "pure" (*pabitro*), "cold" (*ThaanDaa*), and "white" (*shaadaa*). They said that they collect hail in a bottle and drink the water as soon as the hail melts. Like rain, they said, hail water is a "special gift from God to us" (*aallaar daaniioy paani*). Later in the discussion, however, they said that when hail starts to fall everyone becomes alarmed. All people, including children, start to pray and read the Kalima, calling on God to stop the hail storm. Men and women, they said, throw large stones and/or grindstones into the courtyard. These women had seen hail stones that weighed as much as one kilogram.

Such remarks and the magical actions described reveal anxiety and fear of hail. Throwing large stones and grindstones is a type of fight-back action. Positive remarks made earlier about

---

*Suzanne Hanchett and Anwar Islam notes

Photo 4-10. A single, empty pot set at the homestead boundary to fend off danger from hail. Homna Subdistrict, 2009 (Photo credit: Suzanne Hanchett)

hail may well have represented an effort to appease, flatter, or fend off hail as a potentially destructive environmental force.

Summary

This chapter discusses some points of Bengali water vocabulary and common perceptions of water's qualities. Certain types of water bodies, large and small, are differently named and differently used. General qualities of water are explained by villagers themselves: heaviness or lightness, color, age, gender, absorptive capacity, clean/unclean, and pure/impure. Qualities related to humoral medicine are wetness and metaphorical or physical "hot"/"cold." All of these perceived qualities influence domestic water uses, especially when children's health is thought to be affected by them. Some ideas about rain and hail are discussed, and assumptions about the danger of speaking about certain things, especially hail. A vocabulary list in Appendix 2 demonstrates patterns of water language by region.

5. Water in the Community, the Home, and the Life Cycle

*"If water is good, everything is good."**

This chapter focuses on domestic water uses and the factors influencing people's access to the water they want and need, in the context of the overall water resources situation. Domestic use of water does not receive much attention in policy circles. Cooking, bathing, collecting drinking water, laundry, and other routine duties of women are assumed to continue apace despite any changes in water resource management and in all seasons. As Shamim and Salahuddin (1994) point out, women's work is "not seasonal." Whatever changes occur, there is a cost in human time and energy, but this cost is almost never considered when water regimes are altered.

Uses of Domestic Supply Water

In addition to drinking and cooking, domestic water uses include bathing of both people and animals, tooth cleaning, washing of foods, dishes, and cooking utensils, washing clothes, post-defecation washing, and cleaning the house, courtyard, or latrine.

---

*Comment of a Bera Subdistrict woman during a 2009 interview session on allergies and skin diseases

Domestic supply water supply also is used for folk healing, purification before Islamic prayers, Hindu ritual offerings, and in life-cycle rituals (performed at times of birth, marriage, death, for example). Domestic supply also may be used to water kitchen gardens.

Making use of different water sources for different purposes is common practice. We have found this pattern of multiple source use in all of the rural locations we have studied. In a 2002 survey we did for WaterAid Bangladesh in 325 households of ten districts,* typical sources were tube well water (preferred by a large majority for drinking), ponds (for cooking and utensil washing), and rivers or canals (for bathing, but only in some places). Dug wells are less frequently found than in earlier times, but they still are used in some places.** Piped supply water has been introduced in some rural villages. Its convenience makes it popular for all uses whenever it is available.

A 2009 household survey in villages of nine arsenic-affected districts found people using between 2.3 and 3 different types of water sources on average. Those using arsenic removal filters generally used a larger number of sources. Table 5-1 presents information on a control group of survey households, which were not covered by any arsenic mitigation program.

We have found the pattern of multiple source use even in urban areas. Townspeople often have access to some ponds and tube wells as well as indoor piped supply water. Among poor households and in urban squatter settlements (*bastis*), however, almost no homes have indoor plumbing even now. So standpipes, tube wells, and nearby water bodies or canals are likely to be the

*Districts covered by this survey were Bhola, Chapai Nawabganj, Chittagong, Dhaka, Gaibandha, Rajshahi, Rangamati, Shariatpur, Sirajganj, and Tangail Dhaka locations were urban squatter settlements (*bastis*).

**Government and donor programs do not favor dug wells because their water frequently has a high bacterial content.

Photo 5-1. House next to a pond. Barguna District, 1998
(Photo credit: Suzanne Hanchett)

Photo 5-2. Washing cooking utensils at the family pond
(Photo credit: Laila Rahman)

Photo 5-3. Home-made water filter, called *chaari*. Bera Subdistrict, 2009 (Photo credit: Suzanne Hanchett)

Photo 5-4. Household water filter purchased in the market (Photo credit: Suzanne Hanchett)

Table 5-1. Domestic Water Sources Bangladesh Households Located in Arsenic-affected Areas but Not Covered by an Arsenic Related Program (Multiple Responses)*	
Average Number of Different Sources Mentioned = 2.3	
Specific Water Sources Used	Percentages (n=500)
Shallow tube well: not tested for arsenic	66.6
Pond/Lake/Tank	65.4
Shallow tube well: red-marked (arsenic present)	27.2
River/Canal/Marsh (*bil*)	18.0
Shallow tube well: green-marked (arsenic free)	16.8
Rain water	16.4
Deep tube well (approx. 300-500+ feet)**	14.0
Dug well or ring well	3.6
Household filter, general purpose: purchased or home-made	1.6
Household arsenic-removal filter	0.6/0.2
Community arsenic-removal filter (Sidko plant)	0.4

*Survey conducted by Pathways Consulting Services Ltd., Feb. to May 2009, in 16 subdistricts of nine districts (Brahmanbaria, Chuadanga, Comilla, Dhaka, Madaripur, Meherpur, Narail, Pabna, and Sunamganj): Control group households only

**Deep tube wells were found rarely in areas covered by the UNICEF arsenic mitigation project, called DART, which was being implemented in the areas of study.

Photo 5-5. Women bathing and filling their water pots in a village pond. Muradnagar Subdistrict (Photo credit: Shireen Akhter)

Photo 5-6. Collecting water from a tube well inside the homestead compound (Photo credit: Laila Rahman)

principal domestic water supply sources for them. Rain water is used in most, but not all, places during the rainy season.

In a 1998 survey of 1040 households in six urban areas we found tube well water to be the preferred drinking source for 95-100 percent of households.* A preference for cooking with pond water ranged from 50 to 96 percent.** In arsenic-affected areas in recent years the government has subsidized installation of deep tube wells (300-1000 ft. deep). These are a highly valued source of arsenic free water but expensive to install. Occasionally, NGOs (nongovernmental or nonprofit organizations) also provide them.

Women, as the principal managers of household water use, give serious thought to their domestic water arrangements, carefully selecting the sources they consider best for specific purposes. Reasons for using any particular water source include convenience and practical necessity, or availability of the source. "Availability" can be determined by social relationships and hierarchy at least as much as by physical distance. Three case studies (5-1, 5-2, and 5-3) below demonstrate the types of uses women make of different water sources.

Effects of water types on the flavor and color of cooked food are considered quite important. If tube well water has much iron in it, for example, it is thought to spoil the taste and color of rice and other foods, so is not always considered to be suitable for cooking. Further considerations include perceived qualities of water, such as "hot/cold," "pure/impure," "clean/dirty," "wetness," plus others which were explained earlier in Chapters 2 and 4.

Increasing numbers of economically solvent households in rural areas are making use of various types of water purification technologies, in order to make their water suitable for drinking.

---

*Patuakhali, Bauphal, Bamna, Noakhali, Chatkhil, and Feni (Government of Bangladesh and Danida 1998)

** Source: Government of Bangladesh and Danida 1998, Summary Vol. 1

These include a variety of filters to remove bacteria and/or arsenic. Some filters are home-made. Others are purchased in rural markets. Public understanding of filters is weak, and there is little standardization.

A significant gap remains in the case of salinity, which affects water supply in the southern coastal belt areas. Salinity is a difficult problem to solve, and few are coping well with it. For people living in Shyamnagar Subdistrict of Satkhira, climate change is not just something they read about in newspapers but is a reality that affects their everyday lives. In 2012, Tofazzel Hossain Monju spoke with Selina Akhter, a 33-year-old woman living in one of the most affected areas, Gabura Union. "There used to be many seasons here," she said. "But now there are only two. The summer is very hot, and the winter is extremely cold. There isn't enough rain, so we have a big problem getting safe water." Like other southwestern locations, this area has a serious problem with salinity in drinking water.*

Different water uses may compete with each other. For example, using soap for personal bathing or laundry is said to make pond-cultured fish taste soapy. Soap also was mentioned by one women's focus group as making water unsuitable for use in folk healing. "Good water is tube well water," said women joined in a 2009 Bera group discussion. They all agreed that, "A good pond (*maiThaal*) is one where people do not bathe. We can take that kind of water to a folk healer (*kobiraaj*)," they said, "and the healer can blow prayers into the water to cure diseases such as stomach pain, headache, vomiting, or if someone has bad eyes (*shaapa baathaas laage*)."**

---

*October 2012 report by Tofazzel Hossain Monju

**A woman's focus group in Ruppur Union, Bera Upazila, Pabna District, July 2009. Before use of tube wells became widespread, people collected water for domestic use from dug wells, reserved ponds, and *dighi*s. Nowadays, however, reserved ponds and *dighis* are not common; but we have reports of some in Hatiya Subdistrict of Noakhali District. They are said to be "blocked" (*abodha*)—that is

Moving around the countryside, our team has encountered many situations in which pond owners cultivating fish tried unsuccessfully to prevent neighbors from using their ponds for bathing and dish-washing with soap. A number of people not owning ponds confessed to using soap secretly while bathing in others' ponds. One woman in Pabna District described what she called the "bad behavior" of one pond owner. "They use language like this: 'You cannot use soap in my pond (*maiThaal*). If you want to use soap, dig your own pond and use it. If you use soap, you will eat your son's head or, your husband's head', which is a curse."*

Commercial pond fish culture and drinking water are mutually exclusive options. The chicken manure used to feed fish makes pond water unpalatable for drinking, but the water still is considered suitable for cooking in most places. This conflict has led to failure of some initially effective pond sand filter systems, many of which have been installed as alternative drinking water sources in arsenic affected areas (See Chapter 6).

---

unavailable for routine public use, and thus sufficiently pure for use by healers (*gunin*). (Shireen Akhter report, 2007)

*Pabna District, July 2009. This restriction is imposed by pond owners for several reasons. Not only does soap make water turbid and unsuitable for fish culture but there also is an increasingly restrictive sense of ownership of resources such as ponds and orchards. In earlier times, such resources were generally available for common use by poor and landless people, as mentioned in Chapter 1. A long tradition of allowing pond water use by the landless people in Bangladesh is changing, however. Access to pond water is a great problem for rural poor women nowadays. The poor families tend not to have their own ponds. In earlier times, they had access to ponds owned by the rich and middle class families of their communities. But these now are mostly used for fish culture; and the use of pond water is now very limited.

Seasonal Variation in Domestic Water Sources and Uses

In this climate, heavy monsoon rains (June to October) alternate with periods of almost no rain. Of the six named Bengali seasons, four bring dramatic changes in water availability and uses: winter *(shiitkaal*), spring, (*boshontokaal*), summer (*griishokaal* or *griishmo)*, and the rainy season (*borshakaal).* Winter, spring and summer together form a dry season, when surface water, including small rivers, canals, and ponds, becomes increasingly scarce. Small rivers and canals may run dry for part of the year. In the rainy season, surface water is replenished and refreshed. Case Study 5-2 describes a typical annual cycle of water availability and use in Laksham Subdistrict.

Seasonal water scarcity, while considered normal, puts extra pressure on a housewife's time. The introduction of the tube well has helped to reduce this pressure, although tube wells also may provide insufficient water at the end of the dry season, especially in low water table areas. And the use of tube wells for irrigation may deplete the supply available for domestic uses. Pond water normally used for domestic purposes may be diverted to agricultural fields in the dry season. Seasonal water scarcity not only affects women's water collection routines. It also affects social life, as the necessity to share a reduced number of water sources creates social tension.

Gender and Domestic Water Management

Domestic water use and management is mostly a woman's responsibility in Bangladesh homes, but we have found regional differences in water-related gender roles. In a 1998-99 baseline study of water and sanitation in five districts, we found that men were involved in carrying water and other tasks to a greater extent in some areas than in others. In three southern and western towns (Patuakhali, Bamna, and Bauphal), our household survey found 23.5 percent with males having the principal water-collection responsibility. In the southeast (Noakhali and Chatkhil), however, males had this responsibility in only 4.5 percent of surveyed households (See Figure 5-1).

Two reasons for this difference were (a) the more conservative nature of society in the southeast, and (b) the lesser availability of deep tube wells in the south and west. In the areas where it is necessary to travel farther to get tube well water, males are more likely to perform this duty to protect women from violation of purdah norms and any type of harassment by men.

Regional differences are less evident in the case of gender assignment of responsibility for household water storage. (See Figure 5-3) Our survey found that women took on this responsibility in 94 percent to 98 percent in all regions. Pots used for carrying and storage of water are called *kolshi*.

Water storage methods vary by region, but more and more households are shifting away from the use of old fashioned clay pots and using more aluminum or plastic ones. The clay pots allow for evaporation and thus cool down stored water during the hot season, and some people have mentioned to us that water stored in other types of pots is less cool and refreshing. Large clay pots are used to store water for drinking and other purposes (See Photo 5-8).

Sharing Water Sources

Reduced availability of ponds creates tension and conflict. Nonetheless, there is strong moral pressure on individual owners to share drinking water sources with neighbors, whether they are related or not.

Extended kin groups share residential compounds, or homesteads (*baaRi*), in most regions; and the households of such homesteads tend to share whatever tube wells and commonly owned ponds their members may own. However, there are sugnificant differences in the size and organization of *baaRi* compounds from one region to another.* In the southeastern districts, such as Noakhali, Feni, and Lakshmipur, the compoounds tend to be large, having 15, 20 or even more distinct households. In other areas the compounds are smaller, usually just three to five households, each typically headed by a brother from the same nuclear family or a "cousin-brother" related through the patriline.

The larger *baaRis* of the southeast seem to be more self-contained and self-sufficient than the smaller ones of other Bangladesh regions. They also appear to be proud of their separateness. They seem less inclined to share water with people from other compounds than are people living in smaller *baaRis*. This fact can lead to under-utilization of expensive equipment, such as deep tube wells or community-level arsenic filtration plants.

A positive aspect of the southeastern settlement pattern is in greater water security for the poor. Poorer households in large compounds seem to have more security of access to safe water than do poor people in areas with smaller homesteads.

There are socioeconomic class differences in access to all types of water sources in all regions. Rural areas have large populations (average 40 percent) of poor people, whose access to tube wells and ponds tends to be less secure than others'. Although a few have their own sources, poor households are frequently dependent on neighbors' water sources to meet their domestic water needs. Some have satisfactory arrangements with relatives or other neighbors, but water sharing relationships can be disrupted if there are quarrels about other matters.

Most poor families, therefore, are vulnerable to losing access to the domestic supply water they need. Clear statements from multiple study areas demonstrate this point. Some typical comments were these:

---

*A household usually is defined by sharing a common stove, i.e., eating meals together. In some places the word "stove" (*chula*) is used to mean household.

1) "Those who go to get others' tube well water, they search around to avoid embarrassment and shame, to avoid being asked embarrassing questions. They may resort to stealing water to avoid socially awkward encounters." "Those having no tube well openly go to a nearby *baaRi*. But problems may arise concerning children, land, fish, anything. This will affect their access to the tube well water. They will have to go to a different tube well or a pond."*

2) A poor woman in Bera Subdistrict told us in 2009, "When we go to fetch water from other people's houses, they become angry and say, 'Don't you feel ashamed taking water? If our tube well gets damaged, who will fix it?' But we do not reply, because we need water. If they say something, we remain silent." ** This is a common situation, because deep tube wells providing arsenic-free water are installed mostly in or near the homesteads of elite families, who are responsible for their upkeep.***

Water sharing problems can occur at any level of society. Even within a homestead compound, social conflicts may result in one household's cutting off another household's water access. If the water source is a tube well, the owner or caretaker may remove the pump handle to prevent others from using the equipment and put it back on only when she herself needs to get water.

*Laksham Subdistrict tea stall discussion, July 2009

**Tube wells installed by a government agency in rural areas are located near to specific residences, and the members of those households are designated formally or informally as caretakers for the facilities.

***Tube wells installed by a government agency in rural areas are located near to specific residences, and the members of those households are designated formally or informally as caretakers for the facilities.

Photo 5-7. Girls carrying water in vessels called *kolshi*. Barisal District, 2000 (Photo credit: Laila Rahman)

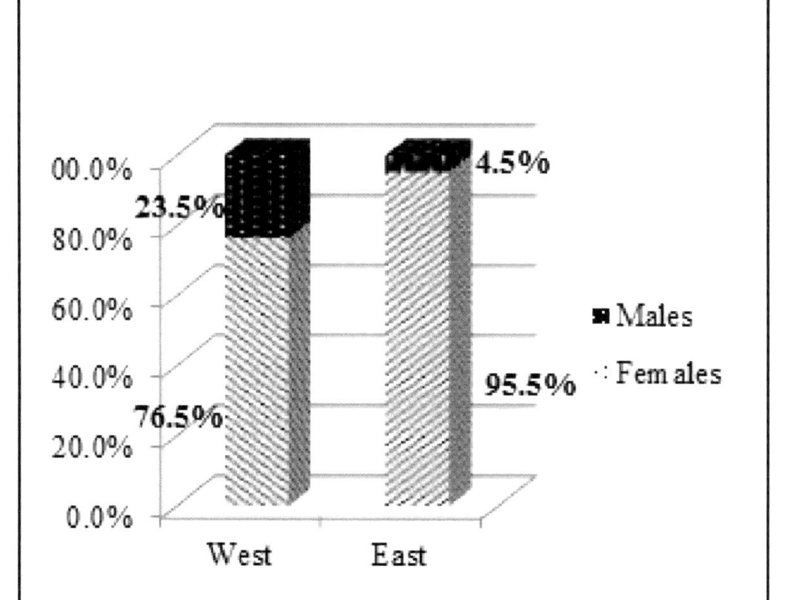

Figure 5-1. Drinking Water Collection by Men or Women in Five Towns (n=656 households)
*Western towns: Patuakhali, Bauphal, Bamna; Eastern towns: Noakhali, Chatkhil.
Source of Tables 5-1, 5-2, and 5-3: 1997-98 survey by Pathways Consulting Services Ltd. for DPHE-Danida baseline study (Government of Bangladesh and Danida 1997-1998)

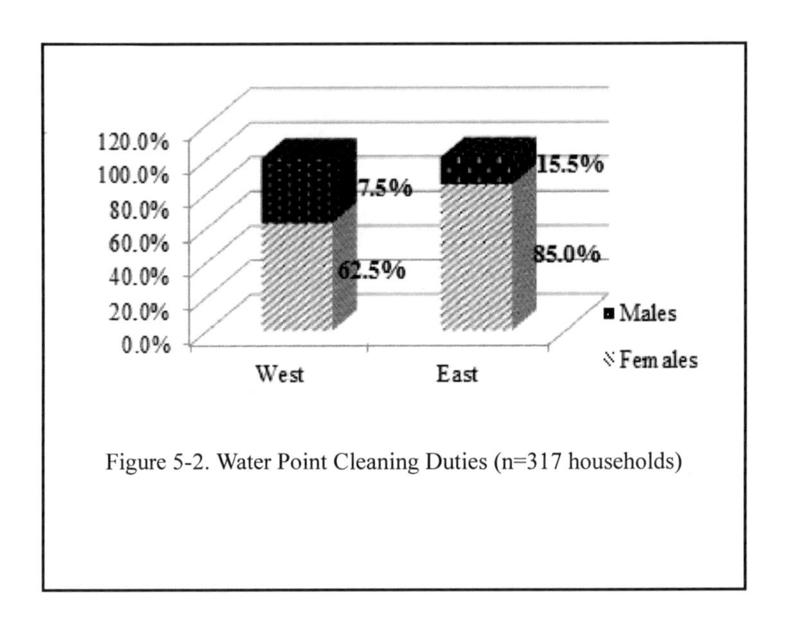

Figure 5-2. Water Point Cleaning Duties (n=317 households)

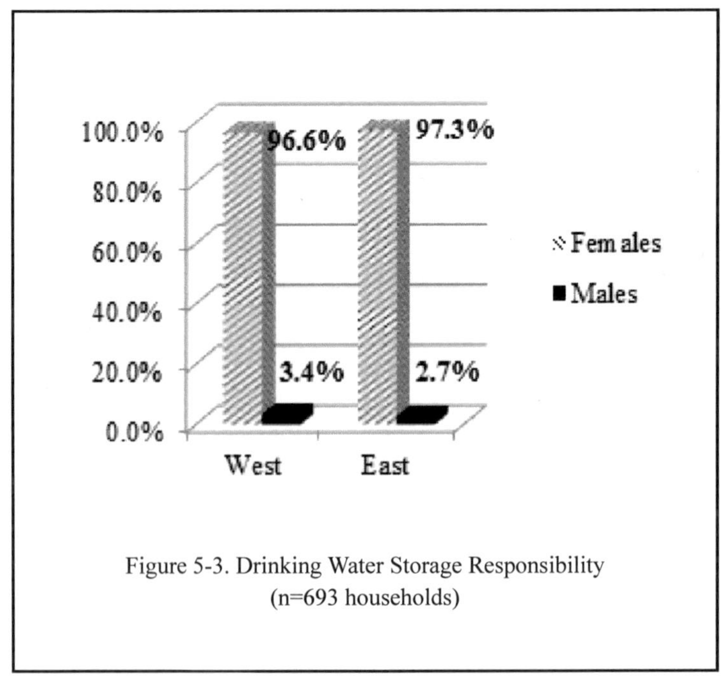

Figure 5-3. Drinking Water Storage Responsibility
(n=693 households)

In one large Laksham homestead, having 20 related households, quarrels over sharing pond water were resolved by building extra steps down to the *baaRi's* pond.* The second wife of a deceased union council member, Fatema Khatun lives with her married sons in this large homestead, where there are four tube wells and three ponds. She explained that there was simply too much pressure on the steps down into the main pond. "I have four sons," she said, "and now they are all married. So there are four different households living in our *baaRi*. Everyone was using the one stairway (*ghaaT*) that led down to the pond. The excessive pressure caused the steps to break. I asked them all to fix the *ghaaT*, but no one came forward. This caused a huge quarrel between my sons' wives and me. As I am an old woman, I could not fix it myself. I became so angry, that I totally broke up the steps. After a few days my sons and I sat together. We decided to build separate stairways down to the pond. So now we have five sets of stairs in one pond!"

Although socially tense situations are inevitable and can be found anywhere, in most places visited we also have met some generous people who agree to let neighbors use their privately owned tube wells and/or ponds. In Bera, for example, in 2009 we met a college principal and his wife who went to great lengths to share use of their perennial pond with neighbors of all social classes. They cultured fish in the pond but did not feed chicken manure to the fish; so the water was potable. Their pond water was said to be used by at least 40 other households for drinking and cooking. The only condition was that no one except the owner's family was allowed to bathe in the pond.

---

*Report by Kazi Rozana Akhter, Laksham Subdistrict, Comilla District, July 2009

Whatever the relationships among the concerned families, the giving and taking of water on a daily basis is fraught with social meanings. A woman of one poor Bera neighborhood that has its own pond told us in 2009 that her community prefers to share the water with other poor people rather than with more affluent women. She said they find poor people to be more respectful and appreciative of their gift. Whenever non-poor women come to use the pond in this poor neighborhood, she remarked bitterly, these outside women criticize their poorer, pond-owning neighbors.

Although many poor people have told us over the years that they must accept abuse in order to get access to domestic water sources, some people choose a less healthy but more socially peaceful alternative, namely, the arsenic-contaminated tube well.

One 50-year-old woman in Pabna District explained her decision in blunt language. "I will not tolerate any more screaming fights about water," she said. "So I recently installed a tube well for our family's use. We did not have the water tested for arsenic. Maybe we will be affected by arsenic diseases, but we have no alternative. If we go to fetch water from those having arsenic-free water sources, they dislike us. They show anger. I ask myself, why are they so proud about a simple thing like water? If we go to fetch water from a neighbor's pond in a better-off part of the village, the owner becomes proud, shows her power. So now I have my own tube well. If I die, so be it. If community people had better relationships and understanding, people would not have to suffer like this."*

Many such conflicts occur despite the widespread and strongly positive value placed on water-sharing. This value is based on Islamic tradition (Hadith or Sunna), which was codified in the 9th century. As many Bangladeshis say, "sharing *(bhaag-koraa)* water is good." No one will disagree with this statement. But in our conversations and observations we have found numerous

*Pabna District, June-July 2009

cases like those just mentioned. One all-male Comilla focus group, re-thinking the water-sharing ideal, came up with the idea that it would be good if social and commercial situations were managed differently. They agreed that, "There are two different social contexts of water use. One is normal domestic use, especially drinking. Every member of the society should collect water for domestic use without any interruptions. In a legal sense, anyone could be the owner of the source. But he or she should not stop anyone from taking drinking water. In regard to the second type of water use, however–for irrigation, fish culture, or industrial production—there must be some rules for use, and users should comply with those rules."*

Case Study 5.1. Annual Cycle of Water Availability and Use

Location: A village in Laksham Subdistrict, Comilla District, Information source: Group discussion with six women Date: 7 July 2009, facilitated by Ms. Roksana (Pathways Consulting Services Ltd.). Reference is to months of the Bengali calendar.

Baishaakh - Joistho [mid-April to mid-June, Summer Season]
These months make up the peak summer season. "This is a problematic time for us. The pond water volume decreases. The rain is not regular. At this time we use more tube well water. Most washing is done with tube well water. We also use Sidko [neighborhood arse-nic removal filter] water for many purposes." The women agreed that, "We feel pressured by the water shortage. We give high importance to water-related works. We instruct family members not to waste water because of the seasonal shortage."

Ashar - Shrabon [mid-June to mid-August, Monsoon Season]
Three women said, "This is the rainy season. We call it the 'water

*June 2009, Laksham Subdistrict, Comilla District, Tofazzel Hossain Monju report

(*paani*) season'. This is a time of excess water, so we need to think about drainage and water collection or conservation. Pond and tube well water are both plentiful. The rain water we get easily is used for cooking, drinking and bathing. Sometimes we seek help from other family members in collecting rain water. We also seek their help with draining water out of the courtyard. Pond water becomes dirty, so people don't want to use it for cooking. If there is not enough water, we suffer an acute crisis, so the rainy season is crucial to our water situation."

Bhadro - Ashin [mid-August to mid-October, Autumn Season] "Bhadro -Ashin is a very good season for water availability and use. The previous season's rain water remains in our ponds. There is no demand for irrigation water, so we have a normal supply of water for domestic purposes. We smoothly plan our domestic water uses and experience no special problems."

Karthik - Agrahon [mid-October to mid-December, Late Autumn] "In this season, the weather is becoming cooler, so people's water use pattern changes. We are on the way from hot to cold weather. In hot weather, there is a demand for more water for drinking and bathing. Karthik - Agrahon is not a total shift of weather, just mid-way. We avoid water sources that are too 'cold'. For example, in summer we sometimes bathe with Sidko water, but we never do in this season."

Poush - Maagh [mid-December to mid-February, Winter Season] "It is the cold season, so we prefer to use warm water. Our water use pattern changes to some extent. The availability of water is normal, but people want to avoid 'cold' water. We identify some ponds as colder than others. In winter we avoid bathing and washing in the early morning. We start only after the sun rises. Old people and children bathe with heated water."

Falgun - Chaitro [February to mid-April, Spring Season]

"This is the beginning of summer. It is also the beginning of our seasonal water crisis. We begin to think about water shortages. If it is too hot and there is no rain, we feel that the water shortage will be acute. Our men plan for irrigation. If there is much land to be irrigated, then more water will be required. We become cautious about how we use water in our homes. If our normal ponds and tube wells run out of water, then we will have to go further away to collect water from remote ponds and tube wells."

Case Study 5-2. Fatema's Daily Water Use Diary

Fatema is a married woman, age 32, educated up to Class VII, living in a five-member household in Laksham Subdistrict. Report by Ms. Roksana (Pathways Consulting Services Ltd., 2009).

Fatema said, "Yesterday I rose from bed early in the morning and washed my hands and face with tube well water. The tube well was red-marked [indicating presence of arsenic]. A couple of years ago, people like you came, and after testing the tube well, they marked it red. Yesterday I cooked with rain water. This rain water was collected two days ago from the roof run-off into a cooking pot. We store rain water with our available pots and pans. If the rain is heavy, then it is a good chance to collect and store water.

"Between 7 and 8 o'clock in the morning, I washed last night's dishes and cooking things, using mainly pond water.

"At breakfast, I poured drinking water from our neighborhood Sidko plant into a jug and kept it on the table. All other family members poured themselves some water from the jug and drank it.

*"At 10 o'clock in the morning, I went to the Sidko plant and collected filtered water for drinking. The Sidko is not far from my house. I can carry water in a medium-sized water pot (*kolshi*). The Sidko is just 100 feet away from my house.*

"At noon, I brought some clothes to the pond and washed them with pond water. After washing the clothes I took a bath with pond water and washed my sari and other clothes with the same water.

"At lunch, I once again kept a water jug on the table. I did not wash the dishes in the pond after we finished lunch. I used tube well water. The reason why I do my morning dishwashing with pond water is that the many dishes being washed with ash and other cleansers would make the tube well platform (kaltalla) too dirty. Actually the morning washing of utensils is heavier than the other two meals. So we do the other two meals' utensils with tube well water.

"I have no little children, so my water use pattern in domestic works is different from those who have kids. All members of my family participate in domestic water collection, and they serve themselves. My husband carries water from the Sidko plant.

"There is not much seasonal variation in my water use routine. In winter there is water in the pond and in the tube well. But in the summer there is some shortage of water in pond and the tube wells. Pond water is used for irrigation [and the underground water table is low]. In the shortage time, we use Sidko water for drinking, cooking and washing. My husband cooperates in collecting water, so we have no problems.

"If water from our large pond is being used for irrigation, then our other, smaller pond is pumped full up with water. Our village is the Union Council Chairman's village, so there have been many good changes in our water situation through his influence."

Case Study 5-3.
Shahana's Daily Water Routine

By Kazi Rozana Akhter, 2009

Shahana, a 38-year-old mother of five living in Homna Subdistrict, Comilla District, is one woman trying to provide her family with arsenic-free water by making long trips each day to collect water from a distant deep tube well. The trip not only takes time, it also forces her to cross boundaries that are socially uncomfortable for a married Muslim woman trying to observe some degree of honor by restricting her movements in the village according to purdah norms. Shahana gets up from sleep very early in the morning and cleans all the family's living rooms, the kitchen, and the courtyard. Then she goes to collect drinking water. She goes before daylight, in order to avoid the crowds and to avoid being seen by men. She says, "I have to strongly maintain purdah while moving around. This is my father-in-law's baaRi *(homestead), and as a wife from a respectable family, I must maintain it until my death." Shahana brings two pitchers of drinking water from the deep tube well and keeps them in the dining room. Sometimes she collects drinking water for her neighbors also. When her neighbor feels sick, Shahana helps her, and vice-versa.*

Every evening, she collects pond water from a large pond far away for cooking purposes. She collects and pours this water in a big earthen pot (motkaa), as her nine-member family needs much cooking water. She keeps the water in a stable condition (thithaano) for about 24 hours without shaking it, to let mud settle to the bottom, to make sure that her cooked curry's color will be nice. Shahana stores cooking water in her kitchen. From this big pot of stored cooking water, she transfers it to a clay pitcher. She also uses pond water for rinsing dishes and spoons before each meal, keeping this water in a white plastic bucket in her dining room. She says, "I use tube well water to clean all the utensils

Photo 5-8. Large clay pots, called *motkaa,* are used to store water in an urban squatter settlement. Dhaka 2001 (Photo credit: Suzanne Hanchett)

Photo 5-9. Woman pouring water out of household storage vessels (Photo credit: Laila Rahman)

Photo 5-10. A poor family's pond with crude steps in disrepair (Photo credit: Shireen Akhter)

and cooking pots after our meal. But I do not use tube well water to clean the dishes and spoons before the meal, because my tube well water has a high level of arsenic."

*Shahana collects tube well water for use in the toilet and for family baths. She says, "I like cleanliness, and I maintain it. Every day before bathing I collect tube well water for use in the latrine and for other hygiene purposes. After bathing in a far-away pond I come home and do ablutions (*oju*) with tube well water before my mid-day prayers."*

Shahana keeps tube well water for toilet use in one large clay pot set in a separate room close to the family's latrine. When going to the toilet, family members carry a small plastic vessel called bodna, which they have filled from that clay pot. Water for use in the latrine is never kept inside the family's living rooms or kitchen. For washing the latrine, she uses tube well water. She cleans her clothes with pond water. (See also Photos 5-11 to 5-16.)

A Hindu-Muslim Difference on Water Pollution

Some of our observations and interviews suggest that there is a difference between Hindu and Muslim views on human pollution of river water. Urination specifically seems to be taken more seriously by Muslims, but we have only limited information on this.

Religious Muslims in Delduar Town, Tangail District, generally believe that urinating in water is a type of sin that God will never forgive until the last day of eternity. This type of sin is called *kabiraa gunna.* One winter morning Anwar Islam observed a Hindu woman and a Muslim woman having a fight on the bank of the River Louhojong. They were shouting aggressively at each other. The Muslim woman accused the Hindu, saying, "Why did you urinate in the water? It is strongly prohibited in our religion.

It is an offense to our religion." The Hindu woman replied that, "In our religion it is allowed to purify ourselves. Urine and water are the same. We believe there are some hidden agents in urine that mix with water to strengthen its purifying power. You can do nothing to me." As they argued back and forth in this manner, a respected old Hindu man of the village stopped the fight by reminding them that, "All of our gods and goddesses instruct us to show the highest level of respect to each other's religion." The women then agreed to stop fighting and left the river bank.*

Kazi Rozana Akhter encountered the same Muslim restriction in Comilla and Chapai Nawabganj Districts, where urination in ponds and cleaning after defecation in ponds are said to be regarded as serious sins among Muslims. During field interviews in these two districts, women told us, "You can do it secretly in the pond while bathing, but Allah sees everything, even our secret acts. If we do it, we pollute the pond water. With this water we do ablution five times a day before Islamic prayers." Another statement was that, "We cook with pond water. It must be clean." Women also mentioned that they do not clean their menstrual cloths in the homestead pond. Two Hindu women in Chapai Nawabgonj told us that, "If we urinate secretly and clean menstrual cloths in the pond, the water goddess will certainly punish us."**

Case Study 5-4. Nokhali District Water Life

By Shireen Akhter

The day-to-day picture of domestic water use is very complicated in Bengali-speaking rural communities. Especially in rural areas, several different sources of water often are used for various

*Report from Anwar Isam on an incident observed in the 1980s in Tangai District

**Kazi Rozana Akhter report, 2009

domestic purposes. Although they may not always have access to their ideal sources, people have strong preferences. This general description is based primarily on my work in Noakhali District between 2005 and 2010. My family home is in a neighboring district (Feni), and I have visited the region regularly since my childhood days.

Bathing and Washing

In Noakhali District the most-used water source in both urban and rural areas is the family pond, most often a rain-fed water body. It is used for all purposes except drinking: that is, for bathing, cooking, washing clothes, cleaning utensils, and washing cattle.

Almost all of the more affluent Muslim residential compounds (*baaRi*) of this area have two ponds. Each is designated for use by different people and for somewhat different purposes. One pond is said to be located "outside," which means just at the compound boundary, usually near the entrance. It is generally used by males and children. A different, "inside" pond, located well within the compound, is used by women.* (See Photos 5-17, 5-18.)

The "outside" pond is usually larger and cleaner than the "inside" pond. The "outside" pond is the preferred place for washing cattle; and the water is considered better for cooking. Men use this pond for bathing, washing some of their own clothes, and also for purification before prayer. The "outside" pond's water is not used for washing kitchen utensils or preparing food for cooking, but in some parts of the Noakhali urban area people pump water from their ponds into their houses. Usually women wash

*This pattern of "inside/outside" ponds also is found in some other southeastern Bangladesh rural districts. "Inside" ponds often are fenced in by bamboo, tin sheets, or cloth, depending on the economic conditions of the household. Some encircle the whole pond (*paaR*), while others shield only the steps (*ghaaT*) leading down to the pond. (Sources: DHV 1998; Shireen Akhter 2007 small group discussions with Noakhali *char* women in Hatiya Subdistrict; and Kazi Rozana Akhter information on Comilla District)

kitchen utensils in the "inside" pond. For cooking, they occasionally send others to collect water from the "outside" pond, as they tend not to have easy access to it.*

Rural and urban areas are crisscrossed by drainage canals that carry off rain water or connect to rivers. People in Noakhali use canal water, if they live near a canal, for bathing, washing clothes, fishing, and other domestic purposes, but not for drinking. Canal water may be polluted with trash, fecal matter from a few "hang latrines"** and waste from small industries. Ponds also may have latrines emptying into them, though there are many fewer since the end of the National Sanitation Campaign in 2006. Many express the view, however, that canal water during the rainy season is not polluted, because it is new, flowing water that comes with abundant seasonal rains.

Men in Noakhali (like men elsewhere in Bangladesh) say that Muslim women are discouraged from using their "outside" ponds because of purdah and their monthly polluted condition. One man told me that, "Women's bodies are beautiful and attractive. So, if a woman takes a bath in an open place, men will be sexually attracted and watch her lustily. He used the word *kudrishTi*, which means "a harmful/sexual look." As purdah rules in Muslim communities forbid women from showing their bodies to unrelated men, women often must use comparatively dirty water from the smaller, "inside" ponds for bathing (In some cases, however, women go to the "outside" ponds at night when nobody can see them.). A young Noakhali wife is advised not to go to a pond when male members of the household or homestead are present.

---

*According to Aziz and Maloney (1985), it is not considered respectable for a young woman even to wash clothes in public places such as river banks, canals or pond access steps, because men might find her body attractive This extreme view, however, is not common in the areas where we have done our studies.

**Hang latrines are ones with seats or squat plates that extend out onto open ground, canals, or water bodies. The structures may have walls and doors, but human feces are not contained in pits or septic tanks.

Women in various parts of Noakhali District say that they like pond water for bathing and cooking because it feels "weightless/light" (*paatlaa*). Pond water that is exposed to large amounts of sunlight and air is said to be less "cold" and not to cause fever or nasal sinus congestion. Pond water, they say, also cleans clothes nicely, especially if there is a large quantity of it. Pond water is thought to clean away all dirt better than tube well water.* Similarly, "Bathing in a pond cleans the body: every pore of the skin and every hair follicle is penetrated, making the body pure according to religious norms."** On the other hand, tube well water, coming from underground, is considered to be "heavy," "cold," and likely to cause fever and other cold-related sickness if used for bathing. Children are thought to be more vulnerable than others to this danger. (This issue was discussed in Chapters 2 and 4.)

Drinking

Most Noakhali people drink only tube well water.*** People love to drink tube well water, because, they say, it is "nice," "cool/cold," germ-free, and tasty. It looks clean and "white," comes from underground, and is a gift from God.**** In those places where

---

*Source: Interviews with both males and females in Noakhali District

**Source: Group discussion in Noakhali *char* "sandbar island," many women from Hatiya Subdistrict

***Two kinds of tube wells are "shallow," taking water from underground sources from 12-30 meters, and "deep," which are usually dug to more than 50 meters. Water from shallow tube wells is far more likely to be contaminated with arsenic than that from deep tube wells in affected areas such as Noakhali.

****The positive term for "cold" (*ThaaNDaa*) in this context is the same word as the negative term used earlier as an explanation of why it is not good to bathe in tube well water. Used in the negative sense, the term refers to illnesses thought to be caused by exposure to cold. Used in the positive sense, it has the

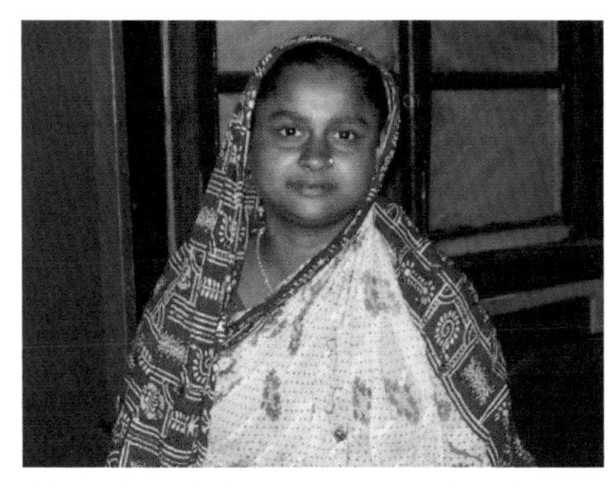

Photo 5-11. Shahana is a housewife in Homna Subdistrict who maintains family honor by observing purdah rules.

Photo 5-12. Shahana's house

Photo 5-13. Shahana travels to a distant tube well to collect arsenic-free water for her family's drinking and cooking needs and stores it carefully in her house.

Photo 5-14. Water for different purposes is kept in different types of vessels. These items are used for drinking water at mealtimes.

Photo 5-15. Large clay pots store extra water for cooking.

Photo 5-16. A pot called ***bodna*** is used in Shahana's household latrine for post-defecation and other washing. This pot never enters the living rooms or the kitchen. (Photo credits: Kazi Rozana Akhter)

Photo 5-17. An “inside pond,” large but with crude steps. Noakhali 2008 (Photo credit: Shireen Akhter)

Photo 5-18. An “outside pond,” intended for men’s exclusive use. Steps are made of concrete. Noakhali 2008 (Photo credit: Shireen Akhter)

no tube well is available people drink pond water after adding alum,* or in a few rare cases, after boiling it.**

Cooking

For cooking, people of Noakhali mostly prefer to use pond water. Women say that it does not change the color or the taste of food. High iron content is common in the tube well water of this region. So tube well water causes problems for cooking. Women of other areas often express the same view) Most say that pulses are best cooked in pond water, because they become soft. They will not soften if cooked in tube well water that has too much salt. And they say that food cooked with pond water remains fresh for a longer time than food cooked with tube well water. In some cases people travel some distance to carry cooking water from large, protected ponds, but typically people use water from their neighborhood or homestead ponds for cooking.

Some changes have come with increasing tube well water usage over the past 20-30 years. People who have their own tube wells occasionally do use tube well water for cooking, but they first let the water sit overnight to allow time for iron oxide contents to settle. Those without access to ponds mostly use tube well water for cooking.***

---

connotation of cooling, soothing, and producing a peaceful feeling. See discussion in Chapter 3.

*Alum (Bengali *fitkiri)* is defined as "any of various double sulfates of a trivalent metal such as aluminum, chromium, or iron and a univalent metal such as potassium or sodium, especially hydrous aluminum potassium sulfate, $AlK(SO_4)_2 \cdot 12H_2O$, widely used in industry as clarifiers, hardeners, and purifiers and medicinally as topical astringents and styptics." (http://www.thefreedictionary.com/alum)

**DHV 1998, DHV 1999; Shireen Akhter interviews in Noakhali *char* areas.

***.United Nations Foundation 2003

Water Storage

The Noakhali District women interviewed are very conscious about in-house methods of preservation of water for different purposes, as are women of other areas. After bringing water from a pond, they store it in at least two separate pots: one for cooking and drinking, and the other for common uses such as cleaning or personal hygiene (post-defecation washing). They never use one pot's water for the other pot's intended purpose. This is a strict observance in households of all socioeconomic levels. Noakhali women and others almost never fail to maintain this separation. If women do not keep their household waters separated, they will be socially criticized. People say that violation of this norm is "hateful," making food "unpleasant to eat" and "distasteful, causing an unpleasant feeling."* (End of Noakhali report)

Purity, Separation, and Water

Like others, Bengali cultural traditions view water as a powerful purifying agent. Because the concern with purity and pollution is especially strong in South Asia, however, purificatory uses of water are pervasive and tend to influence daily life. Purity concerns are evident in post-defecation cleansing (usually washing with water), for example. This personal hygiene procedure is performed only with the left hand. The left hand, therefore, is considered unsuitable for handing things to other people or for bringing food to one's own mouth. This point of etiquette is universally observed among urban and rural populations alike in South Asia.

Other states of purity or impurity have to do with the life cycle. This is an especially urgent matter for women, whose birthing activities and menstrual cycles regularly put them in a strongly

*Portions of this Noakhali report were included in our Five District Baseline Study (Government of Bangladesh and Danida 1997, 1998).

polluted state. "Lost purity can be re-established only by ritual … [and] purity is often a precondition for the performance of rituals of many kinds." (Ortner 2011) The polluted person is considered to be in a disordered condition which is physical, mental and spiritual. A person in such a state may be perceived as dangerous to others, so social separation is part of the experience.

Water is used as a symbolic boundary-marker as people move from one stage of life to another: at occasions such as birth, marriage and death. In this way, Bengali-speaking people resemble all others, but they have their own ways of marking life transitions. While life cycle ceremonies differ among different groups of Muslims, among specific Hindu castes, and between rich and poor, they always include bathing, sprinkling of water, and so on, at status-transition points.*

Restrictions on Women's Access to Water

Women's water usage is restricted at various stages of the life cycle and during their monthly menstrual periods.The most stringent restrictions are those imposed on women between menarche and menopause. When a girl matures, restriction of water use becomes an important part of her life. At the time of menstruation, women are not allowed to touch common water sources such as homestead ponds or tube wells. Nor are they to step on water access places: a pond's raised boundary, the steps leading down to a pond, or a tube well platform. A woman cleans her polluted menstrual cloths with water found outside of the commonly used sources, sometimes using very dirty water from puddles, ditches

*The topics of purity and pollution are thoroughly discussed by Mary Douglas in her 1966 book, *Purity and Danger,* and by Louis Dumont (1970). Arnold Van Gennep's classic work, *The Rites of Passage* (1960), describes common features of such ceremonies world-wide and their social meanings. Ortner (2011) provides a useful overview of the topic.

or drains. These practices put women at risk of skin diseases and genital infections.* (Photo 5-27)

A menstruating woman is spoken of either as being "sick" (*aashusto*) or "impure" (*aapabitro* or *naapaak*). Menstrual blood is regarded as a defiling substance that weakens women and harms others. Physical contact with menstrual blood is thought to destroy men's strength. During menstruation, we were told, women are forbidden to go to a river.

One group of women in Bera told us, "We clean our menstrual cloths in the same places where we urinate. We use tube well or pond (*maiThaal*) water to clean the cloths. But after giving birth, we clean the cloths in an entirely separate place, a place where we do not normally go. And we use separate water to clean them. We cannot go to our ponds for these purposes, because *maiThaal* water is used for cleaning and cooking. And people clean their hands and mouths with this water when they are preparing for prayer (*oju*). If we cleaned menstrual blood, etc. there, the water would be destroyed. It would become "disgusting" (*noshTo*) and "impure" (*naapaak*). These bloods are very nasty. People hate these bloods, and we hide them."**

More restrictions were observed by both Hindu and Muslim families a long time ago than nowadays. Due to changing residential patterns—fewer joint families and more nuclear or extended families—and other factors, menstrual restrictions are now not as strong as they once were.*** There still is plenty of evidence of restrictions, however. For example, K.M.A. Aziz and Clarence

*Some women have mentioned itching in the female organs.

**Shireen Akhter report, 2009

***A nuclear family consists of one or two parents and their offspring. An extended family includes one or more additional relatives, such as a grandparent or an aunt or uncle, but no additional married families. A joint family includes two or more married couples. The typical South Asian joint family is made up of two or more married brothers, their wives, and their children, all sharing a common home and eating together. In general, we have found that such restrictions are less strict in urban areas than in rural villages.

Photo 5-19. A Noakhali woman keeps one pot for use in purifying herself before prayer.

Photo 5-20. Doing the ***oju*** purification before prayer

Photo 5-21. Noakhali woman in her outdoor cooking area, with water-preservation pots. Noakhali, 2008

Photo 5-22. Preservation of cooking and drinking water in a poor family's house. Noakhali, 2008 (Photo credits: Shireen Akhter)

Photo 5-23. Water preservation vessels in the home of a spiritual woman

Photo 5-24. The spiritual woman (in white) going into trance with her daughter sitting near her. The pitcher on the shelf above her had red-colored water in it for participants to use in purifying themselves before the ceremony. Noakhali, 2008

Photo 5-25. Red-colored water distributed to session participants (Photo credits: Shireen Akhter)

Maloney describe a belief that a boat carrying passengers may capsize if a menstruating woman is in it. A menstruating girl commits a sin if she sits in the place where elder family members sit or sleep. After her menstrual period is completed, a woman is required to bathe in a special way before resuming sexual relations with her husband.* (Aziz and Maloney 1985)

A menstruating woman cooks for her family only after she has cleaned herself somehow. According to our information, Muslim women can cook when menstruating. Other menstrual restrictions are known to exist. A menstruating woman cannot touch a cow, lest the milk supply be reduced by her "impurity" *(aapabitro)*. She is told not to go near fruit trees or vegetable growing areas. It is assumed that the trees or vegetable plants will die if she touches them. However, these restrictions are not observed by all women in all places.

According to women interviewed in Noakhali District and elsewhere, after having intercourse a woman is expected to bathe before sunrise. This social norm is found among both Hindus and Muslims. This must be done in the dark, so that nobody can see her "shame." She should not perform any household chores until she has bathed. In Islamic religious scripture, there is a strict law requiring early bathing for both women and men after intercourse. The method of washing is complete, or "head-to-foot." In the case of women the requirement reportedly is more strictly followed. Their ritual purity is especially important, because they are normally responsible for preparation and serving of food. Muslim men and women cannot pray if they do not bathe after sexual intercourse. During the fasting month of Ramadan, if they have intercourse at night before having the *sehri* (food taken before dawn), both of them must bathe, otherwise their fast will not be acceptable according to Islamic law.

---

*"All the world over, not only among the primitive peoples, but also among peoples on a far higher cultural plane, the forms of rites attaching to the first menstruation are similar." (Bhattacharyya 1996)

Photo 5-26. Woman praying at a fertility-granting pond. Laksham Subdistrict, 2009 (Photo credit: Suzanne Hanchett)

Water and Fertility

Some ponds are said to help barren women to conceive. One pond in Laksham Subdistrict is located near Dewan Shaheb, a famous mosque. Women told us that a long time ago this pond (*pukur*) was reserved only for the purpose of taking drinking water. People came from long distances to collect the drinking water, but nowadays the water is not used for drinking. Women wishing for healthy babies now come here. The women come very early in the morning, sit on the pond's steps (*ghaaT*), and do *oju*, the ablution normally done before prayers. Then the women offer a special vow while sitting there. They drink two handfuls of pond water and then dip two times into the pond. Some also pour about a half-kilogam of "raw" [unboiled] milk into the pond. Women who come from distant places change out of their wet clothes right there.[*]

Another pond (*dighi*) visited by childless women in the same subdistrict has a big jackfruit tree growing beside it. The procedure is for women to take a small, baby-sized piece of the fruit and eat it while standing next to the pond. They then drink two handfuls of water from the pond. They dip into the pond just once and leave the place in their wet clothes. After doing this, we were told, many women have become pregnant.[**]

Water and Childbirth

Some ethnographic reports provide detailed information on home birthing practices in Bangladesh and the ideas that motivate these practices. We have conducted some interviews on this topic, and ethnographic reports cited below are based on direct observation of specific cases.

The home birth is an activity in which folk beliefs drive many procedures. Thérèse Blanchet describes women's views: "Reproductive physiology and the cultural world of meanings are not two separate, opposed categories of existence, but one. The

[*]Kazi Rozana Akhter report, Laksham, July 2009

[**]Kazi Rozana Akhter report, July 2009

same values pervade and the same cosmic forces regulate both." (1984:97)

Najma Rizvi (1979:219-220) and Thérèse Blanchet both report that a new mother is said to be in a "raw" (*kaãchaa*), vulnerable state because of having lost blood and other bodily fluids. Her physiological balance is disturbed, and she is susceptible to "cold" for 40-45 days. She needs special foods to protect her against the dangers that this cold pose to her and her infant. Several reports mention that she also needs warmed up or boiled water, because unboiled water is also considered "raw," which exacerbates a physiologically (metaphorically) "cold" state.*

Blanchet's description of a birth in Jamalpur District mentions using "sanctified water" *(paRaa paani)*—water that has had a blessing blown over it by a healer (*kobiraaj*) or a religious leader—to help speed up labor and delivery and fend off evil spirits. "Zori's mother kept feeding *pora pani* (water that has been blown on by someone with special powers) to her daughter to give her strength and remove all evils.... [Waiting for the placenta to be delivered,] they tried a variety of techniques, placing a loose knot-less rope over the upper part of Zori's belly, loosening her sari a bit more... and feeding her some more *pora pani."* (1984:83, 85)

Another ethnographic study, done in northeastern Bangladesh by Kaosar Afsana, also mentions use of this type of water in childbirth. Afsana found that

> [Amulets], *panipora* and herbal roots were understood as conferring spiritual power to birthing women. In Apurbabari, women said that birthing women obtained physical and mental strength from the blessings of the divinity by drinking *panipora. Panipora* was usually sought from the *kobiraaj* [folk healer] to accelerate the

*Rizvi 1979:222; Zeitlyn and Islam 1990.

> process of labour. This water was not always blessed with the Koranic *surah* (verses) recitals. In fact, the *kobiraaj* read some chants from the *Sulemani ketab* (magic book of Prophet Sulaiman) and blew over the water. On occasion, local mantras were also chanted. Birthing women drank this *panipora* with deep faith. *Panipora* was also sprinkled over the head of birthing women to offer blessings. (Afsana 2005:102)

In Comilla District, we also heard about "sanctified water" being used in childbirth.* In this case it was called *shaaran paRaa paani*. Women report that this water has the power to relieve the mother's labor pains. "The Huzur (a religious leader or imam) writes some words from the Holy Quran on a glass plate. At the same time he holds up a glass or bottle of tube well water. He reads some Arabic words (*sura*) and blows on the water three times. The pregnant woman pours some tube well water on the plate and drinks it immediately before giving birth." Other types of special water also may be used to ease a woman's labor. One is *Aab Jumjum* (or *Jamjam*) water from Mecca (See Chapter 2). The water is brought back by a returned pilgrim, one who has made the Hajj. "He or she distributes this water among the neighbors. Elder women or men collect some for use at times of childbirth. The mother reads Kalima,** and then she drinks the water in the name of Almighty God."

Some water use involves what is called "sympathetic magic," helping a mother to relax and open up for the birth.*** When a

*Kazi Rozana Akhter report, 2009. This healing technique, blowing blessings on water, is known by different names in different regions. In Bera we found it was called *jaar-phuk*.

**The Kalima (Kalimah) are books of verses from the Holy Quran and selections from Hadith, compiled to help people understand the fundamentals of Islam, and which are recited and memorized by Muslims around the globe.

***"Sympathetic magic" is a term introduced by the folklore scholar, James

Photo 5-27. Noakhali women at a small water body in agricultural fields during the dry season. This is the type of place where they clean their menstrual cloths. (Photo credit: Shireen Akhter)

new mother is suffering serious labor pain and possibly in danger of dying, an elderly respected person of the family, a man or a

Frazer. The principle is that "like produces like," so images of a wished-for state are used; and "many things are scrupulously avoided because they bear some ... resemblance to others which would really be disastrous." (Frazer 1959:8) There are other loosening-up images involved in home births. For example, Blanchet and Afsana both mention that all the women in the room, including the new mother, loosen their hair, rather than braiding or tying it up in any way. A rope without knots is another such item, mentioned by Blanchet as part of the placenta delivery process. To insiders, of course, such practices are reassuring and perceived as effective. Sympathetic magic is a common feature of folk ritual.

woman, collects a dried Mariam Flower (a flower brought back by a person who has performed the Hajj) and puts it into a new clay pot that has been filled with tube well water. The clay pot is covered with a small clay lid nowadays. (An older women told us that in previous times it was kept under a *paalo*, a type of fish trap made of bamboo). As soon as the flower opens up, or "blooms," inside the water, it is assumed that the mother will get relief from her deadly labor struggle and be able to deliver her baby safely." The mother will get relief (*khaalaash*) from the severe struggle of giving birth. Women consider this struggle period as a dangerous, even deadly time (*jiiban moroneer shomoy*) for a pregnant mother.*

Three waters or mixtures are considered very helpful in cases of difficult childbirth. Mentioned earlier, these may be associated with Muslim holy places, especially Mecca, or at least with a religious leader's blessings. The three types are *Aab Jumjum (*or *Jamjam) paani, Mariam fuleer panni,* and *Shaaran paRaa paani /Huzurer panni. Aab Jumjum paani* is water brought by a person who performs the Hajj pilgrimage in Mecca.

Mariam fuleer paani is a dried, presumably rare flower that grows in Saudi Arabia. Mariam is said to be the name of one very famous and pious Muslim woman who was born long ago in Saudi Arabia.**

Shaaran paRaa paani /Huzurer paani: According to women, this water also has strong force to relieve a mother's labor pains. The Huzur (religious leader or imam) writes some words from the Holy Quran on a glass plate (*bortaan*). At the same time he also gives her a glass or bottle of tube well water. The Huzur reads some *sura* (words in Arabic) and blows on the water three times. This water is called *shaaran paRaa paani*. The pregnant

---

*Kazi Rozana Akhter report, Comilla District, 2009

**Mariam was the name of the sister of Moses, who set him afloat to be found by Pharaoh's daughter, though any connection to this biblical event is unclear.

woman is told to pour some of the tube well water on the plate and to drink it immediately before giving birth.*

Disposal of the placenta plays an important role in the birthing process. Blanchet discusses beliefs about the placenta and its proper disposal, which means deep burial, approximately three feet down, in a safe place. The placenta is viewed as "the source of [a baby's] life," she observes. It is the seat of some life-matter, according to folk notions of anatomy and physiology. (Blanchet 1984:89-90) Afsana quotes one woman who said that, "*Atma* (soul or life) remains inside *naar* [umbilical cord], which binds mother with the baby." (Afsana 2005:93) Both studies mention that the placenta is disposed of very carefully, to avoid its being eaten by scavenging animals, as this would have dire consequences. Because the link between the mother and the placenta, or between the placenta and the baby, is not thought to end at delivery, its capture by a wild animal or a dangerous spirit (*bhut*) is feared. Women assume that if an animal or a *bhut* got the placenta, the mother's milk might stop flowing, the mother might become sterile, or the child might become ill or even die.

According to the women Blanchet lived with, burying the placenta near a pond or a river is done to ensure that the new mother's milk supply will be adequate. (1984:90) In this case, the flow and volume of liquid water are understood to have a positive influence on the mother and child through the medium of the buried placenta. This can be seen as another example of sympathetic magic, although South Asian midwives and new mothers take it literally.

Washing or bathing with water is an essential part of the confinement experience in home births, which still are the large majority. There are different customs and practices, and some follow rules more strictly than others do. According to Blanchet, Muslim women generally spend seven days in the special room

---

*Kazi Rozana Akhter report, Comilla District, 2009

set aside for childbirth. Among Hindus, the duration of confinement varies according to caste.

As Afsana explains, women are considered to be "dangerously polluted" during the first six days after childbirth. She found that, "People feared that touch of *chochi* [polluted] women would cease their cows' ability to give milk and destroy [their] paddy field." New mothers, she observed, were not allowed to touch the family stove during the initial confinement period. A baby, however, was not considered to be as polluted as its mother. The baby's hair, grown in the mother's womb, was understood to be nourished with menstrual blood, so it was "polluted" (*chochi*). But "the baby became completely free of *chochi*, as soon as baby's hair was shaved." (Afsana 2005:76-77)

In Hindu families, the new mother is washed by the midwife immediately after she gives birth, but she will not take a full bath until her period of confinement ends. According to Blanchet, "In Muslim households, the prohibitions are not so strict and a confined woman may bathe if she feels the need to do so.... [When leaving the birthing/confinement room to defecate or urinate,] a confined woman sprinkles her path with water to remove all danger in front of her. Behind her, someone sprinkles again to remove traces of pollution.... Before re-entering the house, she purifies the soles of her feet over a fire and steps inside without turning back." (1984:99-100)

In Dera, we were told that after a woman gives birth her bloody clothes and other cloths are cleaned up by some relatives. "First we clean with river water and later we clean with soap," the women said. "We do not clean these items in ponds, where the water is not moving (*bodho jal*). Those bloody clothes are poison, and people drink and bathe in pond water. But river water is flowing (*cholti jal*). If you wash dirty things in it, it will take everything away. Some clothes are cleaned in a separate place and the water is thrown into another place."*

The baby is bathed and rubbed with mustard oil after it is born and throughout the initial period of confinement, according to other ethnographic reports and our own information.

Afsana describes one birth she observed. After the umbilical cord was tied and severed, "The baby [a boy] was taken out to the courtyard for a wash... The baby was given a bath with soap and stove warmed pond water. It was wrapped in dry old clothes and taken to the sun." This routine was followed for the first seven days after birth. Each morning the baby was put in the sun after his body was massaged with mustard oil. Nearby was kept a bundle of items with the power to protect against dangerous health influences such as "wind" (*baataash*) or evil spirits.** (Afsana 2005:41-42)

Because of concerns about pollution, contact between a new infant and its father is limited during the period of confinement in cases of home birth. According to Afsana, a father will not touch his child at all for the first seven days of its life, and many men actually do not touch their child for much longer. If they do it is not likely to be in public and [touching] will be followed by thorough bathing. (Afsana 2005:100)

After the initial period of confinement (of seven or some other number of days) a full purification ceremony is performed for the infant and mother. Details of one ceremony are in Afsana's book. The infant's head was shaved, and its fingernails and toenails were cut. After this, the baby was bathed. The observed bath was in a pond. The mother also bathed (in a pond in this case), dipping under the water three times "to remove pollution."

---

*Bera Subdistrict women's comments, Shireen Akhter report, July 2009

**The items were wrapped with a piece of fishing net. They included a cow bone, an iron knife, bamboo broomsticks, a branch of the *nishinda* tree [*Vitex negundo*], and a matchbox. (Afsana 2005:42-43)

Her clothes were washed and changed. "Now she was free of pollution and would be allowed to touch the stove." Earlier in the morning, the new mother's bed had been removed from the floor of the birthing room. The whole house and courtyard were cleaned, and household utensils were washed in pond water. All the clothes that had come into contact with the mother and infant were washed thoroughly. "The clothes were boiled with soap and pond water, and washed in the pond." (Afsana 2005:45)

These measures returned the mother to a somewhat normal condition, but for a Muslim woman, "Full purification was not achieved until 40 days after the birth, when she would perform prayers." (Afsana 2005:44-45) For Hindu families the members of the extended patrilineal family -- the father, his parents, his unmarried sisters, and his brothers and their families -- are considered to be polluted for some period of time after a birth (or a death) of one of their members. This means that they will be kept at a distance from others in their communities until the customary period is over. For example, neighbors will not eat meals with them. According to Blanchet (1984), Muslim families do not follow this rule of extended pollution.

The Bera Muslim custom, according to our information, is somewhat different. For about 40 days, a new Muslim mother is not allowed to go to a pond (*maiThaal*). Nor can she touch a tube well, especially not another family's tube well. If the mother of a newborn baby touches it, we were told, she will destroy the purity of the tube well water. Poor women are even more restricted than others, as they do not have their own water sources.*

These various practices are handed down from one generation to another. Many of the women we talked with explained that their elders have established the rules, and they generally follow them when giving birth at home, as most women do. If

---

*Bera Subdistrict report by Shireen Akhter, July 2009

they deliver in hospitals, however, they seem to follow the rules of those institutions instead. (Afsana 2005)

Khadija said, "If a child is born at night, then his or her bath takes place the next morning, around ten o'clock, depending on the season. If the child is born at the hospital, the bath is done within one or two hours."
Mithu added, "The technique of bathing a newborn child is not the same for all. Some mothers bathe the baby on their lap, and some put the child in an open plastic pan."
*Three women said, "Most mothers will not bathe their babies with 'raw' (*kaãchaa*) water from a tube well or pond. They prefer to use tube well water that has been boiled and kept in the sun for a child's bath."*

Group discussion in Laksham
Subdistrict, Comilla District, July 2009
Tofazzel Hossain Monju report

Because a nursing mother's body is believed still to be connected to that of her baby, she needs to be careful to avoid doing things that will build up too much "cold." (Islam 1985, Rizvi 1979) A young child's "coldness" is believed to be very much related to the mother's condition, as one Comilla group explained. If a mother uses cold things, drinks or even uses cold water, wears wet clothes, or takes a bath late in the day, it is assumed to affect her child's health. Children get cold-related sickness very quickly, they said. Because the mother breastfeeds the young child, its mother's cold is assumed to harm her child's health.* If a nursing infant becomes ill, a mother often will take medicine for that illness instead of the child.

*Kazi Rozana Akhter notes, Comilla District, 2009. See also Mahmuda Islam (1985) on the presumed connection between a breastfeeding mother's physiological condition and that of her infant.

Bathing of both mother and infant is done in a way that is assumed to protect both of them from cold-related illness. The procedure was described to us as follows. Pond water is warmed up *(kumkum gorom paani*) before bathing a newborn baby. The baby is given a bath immediately after it is born. Later on the baby is given a bath every alternate day with warm water. The day when a baby is given a bath the mother cannot bathe. Women said, "If both bathe in the same day, then both mother and baby can get "cold" (*ThaanDii*) immediately." Women said they prefer to use pond water to bathe a baby. They collect pond water in the morning and keep it in a bucket in the courtyard for two to three hours in the sunlight. The sun heats it up. This is called *surja paak kora* (literally, "to purify with the sun"). This water is very good for a baby's health, especially to avoid "cold." To avoid cough or "cold" the nursing mother does her own bathing with warm water. If she bathes, she does it before 10 a.m. Then she dries her hair in sunlight. The mother also avoids cold water and cold foods such as banana, coconut, or watermelon. She drinks only warm tube well water.* (Seasonal variation in bathing practice is described in Case Study 5-1.)

A Midwife's Explanation

*"We are a midwife group or caste (*jaat daai*). My mother was a midwife (*daai*), my grandmother also, and most of our family women are* daai. *We are real* daai. *You will not find others like us. My mother-in-law is also a* daai. *I cannot remember how many births I have attended. It may be more than one lakh [100,000]! I am so expert, that if a child is coming out the wrong way, hands- or legs-first, or if a woman is in labor for three or four days, they call me.*

"When a woman has given birth, I cut the umbilical cord and

*Comilla District group discussion with Muslim women, Kazi Rozana Akhter, facilitator, July 2009

bathe the child with hot water and Detol (a disinfectant), if that is the family custom. Some use river water. Some use tube well water. Some do not bathe the child.

"I also help to clean the mother, some with a bath and some without. It is done in a separate place where no one can see them. For about 40 days new mothers are not allowed to use water from the tube well or pond for bathing. New mothers are "polluted," and this is the rule of the society. The menstrual period is treated in a similar way. Women become naapaak *and* aashucho*—"impure"—and they are not allowed to use regular water sources. Instead they use water in places where no one goes.*

"After sunset (the time called shondhaa*) is a dangerous time for women. Supernatural beings roam around at such times. If a woman sleeps, one might come and snatch her baby. Many women lose their children because they are sleeping during* shondhaa. *They go to a* kobiraaj *to protect newborn children. They put a mixture of four metals in the baby's earlobe. If one piece of heated iron is kept by the baby's side, supernatural beings (*jinni*) cannot come and harm the baby. Sometimes the* kobiraaj *or* hujur *blows on the iron and makes it hot, so that the spirits cannot eat the baby. This is called* mullo dosh.*"**

Some Uses of Water in Weddings and Funerals

We have collected very little information on other types of life-cycle ceremonies, but Ellickson's (1972a) ethnographic report provides some interesting information on uses of water in weddings and funerals in one Comilla District village.

Water in Weddings. "Marriage," Ellickson explains, is "the most elaborated rite of passage in rural Muslim Bengali society." The

---

*Report by Shireen Akhter from Bera Subdistrict, July 2009

events she observed "usually went on for a week." She describes the specific steps, which include the formal agreement, the preparation of both bride and groom for the wedding, the wedding itself, the moving back and forth between bride's and groom's homes, and so on. Each includes complicated gift-exchange, eating or feasting, and, of course, some bathing to symbolically mark the status transitions that are occurring. As with birth, good luck items are collected at certain points and shared with others.

One interesting and puzzling feature of these events, as described by Ellickson, is putting ritual items into the family pond (which she calls a "tank" in English). For example, the occasion on which the wedding date, gifts, and other details are set is called paan-phul, literally "betel and flower." It is so named because representatives of the groom's side are given some paan, i.e. betel leaf and nut, which is a normal welcoming gesture, along with a flower (*phul*). Ellickson observed that,

> The *paan* and *phul* were taken back to the home of the groom, where, the next day, his mother added some unhusked rice—the primary food grain of the area—and some durba, a variety of grass used as an element of Hindu sacrifice since Vedic times. All of these she threw in the tank (man-made pond). Almost every homestead had such a tank which supplied water for drinking, cooking, washing, and often irrigation. (Ellickson 1972a:112-113)

The family pond got another offering after the feast in honor of the groom, which Ellickson calls "the bachelor dinner." She watched as "the groom's mother's father took a handful of rice from the final course," which was plain rice and milk:

> [He] squeezed out the milk and laid the rice aside on a cloth. The next day the mother of the groom threw this rice in the tank, as she had previously with the *paan*,

> the flower, the paddy, and the *durba* grass. (Ellickson 1972a:117)

Yet another ritual item was placed in a pond—this time at the bride's family home—after the bride and groom had been formally wed:

> Some paddy and *durba* grass (*dhaan durba*) was placed in the hand of the groom; he dropped it onto a [winnowing tray, called *kulaa*]. The bride did the same. The *dhaan-durba* was later thrown in the tank. (Ellickson 1972a:122)

The meanings of such folk rituals were not explained or discussed by the participants, who merely said these were their "customs." Ellickson, however, struggles to understand why people might throw these items into their family ponds. She dismissed the idea that the items might be offered to spirits in the water; for, as she explains, such spirits generally were said to be "in the huge tanks, not the friendly little household tanks into which the ceremonial items were thrown."* (Ellickson 1972a:130) Her best guess is based on a Hindu custom of immersing clay images in ponds at the end of certain festivals:

> The image is immersed in a tank after its worship ceremony (*pujaa*) and subsequently dissolved, but the deity symbolized by the image lives on. Perhaps the unity of the individuals and groups which I have suggested these ingredients symbolized, lived on after and because of their immersion in the tank. The affinal [marriage-related] ties which they symbolized were

*We assume that the "huge tanks" she mentions are *dighis*.

extremely important to the strength and continuity of the family and the patrilineal kin group." (Ellickson 1972a:130)

If her interpretation accurately portrays folk-symbolism logic, another metaphoric use of water comes to light. The pond full of water, having the power to dissolve whatever goes into it, also has the power to represent people's hopes for a successful merger of two families united by a marriage. Another possibility is suggested by the fact that the pond is receiving rice, which is the quintessential "food." The implication would be that the pond is somehow alive and thus capable of eating. As a quasi-living being, the water body itself might be able to bless the union and the future of the family.*

A Marriage Custom: Reading The Bride's Footprints

Rural people follow this practice when a man arrives home with his new wife. In front of the relatives and others, someone throws water into the courtyard, making the ground muddy. The new bride goes into the house after walking on that mud. She does not wear shoes when she does this. People observe her footprints and predict whether she is a bhaagovoti, a woman who will bring a good, lucky future to the home. The purpose of this is to ensure a good future, good fate.

Report from Delduar Subdistrict,
Tangail District, by Anwar Islam, 2007

*Though non-verbal, ritual acts do express feelings and ideas about what is going on, but the lack of spoken explanation makes it difficult to be sure of just what is being said—or how these customs might have evolved in the first place. Some interpretations, such as these, can be confirmed or disproved only through indirect study methods. These might include literary or folkloric studies that explore the ways that water imagery is used. Methods of interpretation are discussed in Hanchett 1988.

Water in Funerals. Ellickson found that the ritual items found at weddings and other auspicious occasions (*dhaan-durba*, betel, turmeric, henna, mustard oil) were "completely lacking at a funeral." The only common elements were recitation by a religious leader, feeding guests, and a bath. (Ellickson1972a:125-126)

Ellickson was told that for Muslims the final bath is extremely important, as one must be very clean before entering Heaven. Her report includes a description of one female corpse's bath, which was very thorough indeed:

> A kind of raft was made by peeling layers from a banana plant. The body was laid on this in front of the house in which the woman had died, with bamboo doors set up as a screen to cut off the general view. [A woman specialist who had been summoned to bathe the corpse] was allotted a portion of the new white cloth bought for the shroud.* This she tore into a number of small squares. number of small squares. The first wash water was boiled with two kinds of leaves (*nim* and *borui*)** and a piece of thatch from each of the four corners of the house. The hair of the corpse was washed thoroughly with soap and water and was parted into three sections. Each portion of the body received an application of soap and water two or more times. This was no mere ceremonious splashing on of water. After the final washing of each portion, the [woman specialist] took a clean square of white cloth and rubbed it vigorously on the skin of the corpse. Then she showed the cloth to the assembled women so they

*This woman specialist is called a *dhowyani beTi*.

**These are, respectively, *Azidrachta indica* (Bangla *niim*); and *Zizyphus mauritiana* Lamk. (Bangla *boRoi*), English name: Jujube fruit tree.

could see that there was no dirt remaining. (Ellickson 1972a:126)

According to the custom in southeastern Comilla District Muslim families, the water for the funeral bath is heated, and some leaves of the *boRoi* plant are added to the water. This same community also has the custom of emptying a new clay pot of water on a grave at the end of a funeral. The pot is left resting upside-down on top of the grave before mourners depart. This is said to bring peace to the departed soul. (Photo 5-28) In Delduar Subdistrict, Tangail District, another use of water in funerary ritual is reported: At the very end of a Muslim funeral, as people are preparing to leave the burial ground, the Imam throws some water onto the grave.*

Muslims normally bury their dead, but cremation is the norm among Hindu families. One group in Bera Subdistrict explained to us the importance of a river for Hindu funerals. "When a person dies, we take him or her to a riverside pier (*gaaŋ ghaaT* or *shaashaan ghaaT*) for cremation. We do not clean the body at home. We clean him or her at the side of the river. After cremation we throw the ashes into the water. The river (*gaaŋ*) takes away everything quickly."**

Water and Folk Healing

Water appears in numerous health-related folk practices. The most common of these is for a person with religious or healing powers to recite verses or holy words over water, literally called "reading water" (*paani paRaa*), on the assumption that the spiritual value of the words is absorbed by the water. One Tangail

*Comilla information from Kazi Rozana Akhter, and Tangail information is from Anwar Islam.

**Shireen Akhter report, Bera Subdistrict, Pabna, July 2009

Photo 5-28. A Comilla District grave. A clay pot painted red and white was filled with water and emptied over the head end of the grave. (Photo credit: Kazi Rozana Akhter, 2009)

report on folk healers (*kobraaj* or *fokir*/fakir) refers to this practice. "If someone is seriously ill, feeling abdominal pain, he or she goes to a *kobiraaj* or *fakir*. The healer suggests taking some tidal river water while facing toward the west. The water must be collected in a new earthen pot. The *kobiraaj/fakir* recites some blessings and blows three times on the water. The patient then drinks the water."*

*Anwar Islam report, 2007

Water located near a graveyard where a Muslim saint is buried also may be assumed to absorb the holiness of the saintly person and thus to have healing powers. One example we know of is the Kanchanpur Dorga Bari Dighi in Feni District, which was discussed in Chapter 3.[*]

One healing use of water is a widespread treatment for the body rash called *ghaamaachi, ghumaachi* or *ghaambichii.* The condition is defined by small, itchy red pimples covering the body. This condition is believed to be caused by heat; and it is cured by exposing the patient's body to new rain. Comments from Comilla District were these: "*Ghaambichii/ghaamaachii* becomes visible on the hands, back, neck, and chest, in the months of Baishakh [April-May] and Joishtho [May-June] because the weather is so hot at that time. The color of *ghaambichii* is red, and the pimples have fluid inside them. It itches very much. Raindrops are very good for reducing the rash. They also help to cure other skin conditions. The raindrops are very pure and God-gifted. They heal *ghaamaachii* very quickly, but for other conditions healing with rain water takes more time." The effectiveness of rain water in curing the rash is perceived to result from its purity and its coolness, which counteracts the bodily "heat" that is assumed to cause the rash.[**]

Summary

The focus of this chapter is domestic water, which is used for basic drinking, cooking, bathing, and personal hygiene, along with purification, life cycle ceremonies, and folk healing. The chapter reviews seasonal variations in water availability, water sharing, and their implications for social relationships. Use of multiple water sources for different purposes is commonplace.

---

*Shireen Akhter report, 2007

**This is "metaphorical heat," as discussed in Chapter 3.

There are widespread concerns about the “purity” and “pollution” of ponds and other water sources. The southeastern Bangladesh (Noakhali) region’s practice of setting up “inside” and “outside” ponds around homesteads brings out some gender principles relating to water access and use. Aesthetic and health concerns underlie drinking, cooking, and bathing practices. The chapter reviews ethnographic information from this and other studies of ways that water is used in birth, marriage, and death rituals. Some uses of water in folk healing are presented at the end of the chapter.

6. The Arsenic Problem: Institutional Efforts and People's Responses

Bangladesh and West Bengal, India, have the largest combined population in the world endangered by arsenic-contaminated groundwater. (See Maps 6-1 and 6-2.) A 2000 report summed up the severity of the Bangladesh problem: "It is estimated that of the 125 million inhabitants of Bangladesh between 35 million and 77 million are at risk of drinking contaminated water. The scale of this environmental disaster is greater than any seen before; it is beyond the accidents at Bhopal, India, in 1984, and Chernobyl, Ukraine, in 1986." (Smith *et al.* 2000) Nine affected districts of West Bengal, those in the eastern part of the state, have a total population of approximately 43 million. A survey done in 2000 found 34 percent of water samples in these districts to have unsafe levels of arsenic content, as officially defined by the Government of India, more than 50 micrograms per liter. This study estimates the combined at-risk population in West Bengal and Bangladesh to be more than 100 million. (Chowdhury *et al.* 2000) In Bangladesh, one group of experts suggests that arsenic-related cancers will double the number of cancer deaths during the next two or three decades. (van Geen *et al.* 2005)

The problem of arsenic in groundwater became acute because of expanded use of tube wells from the 1960s onward. These water sources became very popular for two reasons. First, the water was mostly free of pathogens, and secondly, they reduced the need to share. For the first time in history, almost every compound could have its own water pump. Organizations such as the World Health Organization and UNICEF vigorously and successfully promoted tube well use, along with government agencies and NGO partners for more than four decades. Neglect of water testing was the subject of a UK lawsuit filed against UNICEF. Eventually the suit was dismissed. (Ravenscroft *et al.* 2009)

The formal acknowledgement of the arsenic problem and subsequent official actions make up a convoluted story. The problem was recognized in West Bengal in the mid-1980s, but the news somehow did not reach Bangladesh officialdom until the end of the 1990s. Reasons for this apparent delay are beyond the scope of the present book, but they deserve study in their own right.

The largest agency working on the arsenic problem between 2000 and 2006 was the Bangladesh Arsenic Mitigation and Water Supply Project (BAMWSP). BAMWSP started up in 1998 with responsibility for arsenic mitigation in 70 percent of the 269 most affected subdistricts, covering a population of more than 46 million.* The World Bank authorized 44.4 million U.S. dollars for this effort, which ended in 2006.** In 2001, UNICEF took responsibility for action research and blanket screening in approximately 45 subdistricts and mitigation services in 20 others. Other places were covered by various NGOs and medical research institutions or projects. In 2004 the Government of Bangladesh developed a National Policy for Arsenic Mitigation.

The massive push to determine the scale of the arsenic problem and alert the public ended around 2005 or 2006. The water of tube wells mostly had been tested. The most seriously affected

---

*These 269 Bangladesh subdistricts represent 53 percent of the nation's total of 507

**The World Bank required that BAMWSP be established outside of the system of governmental agencies, although it was doing the government's work. Eventually it was formally connected with the Department of Public Health Engineering.

areas had been identified. People started to become familiar with the word "arsenic," though misunderstandings were (and still are) common.

Some arsenic mitigation activities have continued or led on to others. UNICEF has continued to work on the arsenic problem through some special projects and its ongoing water and sanitation program. But BAMWSP has not continued. Nor has it led to new ideas and programs on the anticipated scale. There is very little to show for the huge monetary investment except for the tube well screening database and approximately 9300 deep tube wells* installed mostly during its final two years. (World Bank 2007)

Many mistakes were made during the initial period of arsenic-related activity in Bangladesh. Too much money was given out too quickly by the biggest donor, the World Bank, without any clear consensus on best practices. There was too much bureaucratic bargaining, and not enough program planning or monitoring. Donors now have largely withdrawn, and the government has yet to make a financial commitment to fund long-term, ongoing services. In the majority of affected regions, for example, there still is no convenient or affordable way to get domestic water tested for arsenic content.

Almost all work to date has been done through short-term project activities, and the coming and going of short-term staff has been confusing to the public. The general public seems to expect and assume that higher-ups are providing needed services. The concept of a "project" is not familiar to many villagers, so confusion is commonplace even now. Furthermore, there has been very little consistency, continuity or coordination of arsenic mitigation activities, whatever the public's expectations. With a few important exceptions, the pattern has been to rush forward

*"Deep" aquifers tapped by these wells tend to have arsenic-free water. They range from 200 to 900 feet in depth.

with some new ideas, try them for a while, and then to rush away without much follow-up or monitoring of outcomes.

Arsenic related health problems have affected poorer people to a far greater extent than others. The reasons for this are not fully understood. Differences in nutritional status presumably account, at least partially, for this difference.

Even now the affected public has not yet been properly informed about the arsenic problem. In 2009, we found rural people, even educated people, in targeted project areas, to be inadequately informed about arsenic risks and safe water options. National newspapers, however, do continue to cover the arsenic problem.

The Cultural Dimension: Coping with Arsenic and Arsenic-related Illnesses

We have been observing the reactions of the Bangladesh public to news of the arsenic problem since the late 1990s in the course of working with several different water improvement projects. Many puzzling situations have come to our attention. For example, in 2006 we visited a compound in Muradnagar Subdistrict with a serious arsenic problem, where the elder son of a large joint family proudly showed us a rain water harvesting unit the family had set up. He was a village doctor. He told us that the family gets arsenic-free water from this source for nine months of each year. His wife showed us colorful posters explaining the importance of drinking arsenic-free water. As the mother of an 8-month old baby, she was very concerned about the danger that arsenic posed. His elderly mother, however, was cooking the meals for this ten-person family with arsenic-contaminated water from the compound's own tube well rather than making the trek to procure water from an arsenic-safe well they said was "some distance" away from the home. The elderly woman said that she had not participated in any of the meetings that had been organized to inform villagers about the arsenic problem, because her husband did not allow her to go outside for such meetings. This case shows that differences in age, gender, and perhaps also an old woman's lack of participation in public discussion ultimately can compromise attempts to improve family practices. (See Case Study 6-1)

The rural people most directly affected by this problem themselves generally have had complicated responses, as several studies have demonstrated by now. Some people, especially the more educated and affluent, have stopped drinking arsenic contaminated water. Many are still confused, as new wells, filters, and other options appear in their villages, often without enough background information, or through projects which end abruptly. Some have tried to shift to safer water sources but eventually have returned to drinking more convenient, arsenic-contaminated tube well water.

In the early stages of the arsenic mitigation projects, we found that participants in several focus groups thought of arsenic as a new kind of water-borne disease, which could be fixed by boiling water. Hand-pumped tube well spouts had been painted red, to indicate the presence of arsenic in the water, or green, to indicate safety. (See Photos 1-6 and 1-7.) But perceptions of the red or green color markings were not always clear. Perhaps the use of red or green made sense to planners familiar with traffic signals that signaled "Stop" (red) or "Go" (green), but in rural areas they are less familiar. A further source of confusion was that red is also the color of water with too much iron, and which leaves a red residue in water containers, on clothing, and in some cooked food. So some people thought that the red color signified presence of iron, which was a commonly recognized problem with some tube well water. (Hanchett *et al.* 2002) Folk beliefs influence responses to new technology, as discussed earlier, in Chapter 3.

We found similar concerns associated with household arsenic removal filters when they were first introduced. In a 2009 study of people's responses to newly introduced arsenic-removal filters. We found the water from all types of filters to be appreciated for its good taste, cleanliness, clarity, freedom from iron, safety, and other positive qualities. The filtered water, however, also was said to be "too cold in the winter and too hot in the summer." Beliefs about cold causing illnesses are so strong that many mothers do not give the filtered water to their children when they are sick. (UNICEF 2009)

An early study in six affected Bangladesh towns summed up the public's initial response to the arsenic problem shortly after an awareness raising campaign had reached them. "Focus group interviews showed that the arsenic crisis is sorely testing the public's ability to understand and accept new water use messages. [The messages are not only] unfamiliar, but also they contradict what has been conventional wisdom about "safe water" for more than a decade. ...Perceptions of the crisis itself turned out to have some surprising twists....Many of the affected people think that the earth or tube wells have changed, or that some new kinds of diseases have come into existence. Some unknown number of people think that the problem is in the tube wells: that the tube wells have deteriorated somehow, causing arsenic to contaminate the water." (Hanchett *et al.* 2000:70-71)

Meanwhile, shops selling tube well equipment started painting their pumps blue, which some people took as evidence that they would produce safe water. (Photo 6-1)

Photo 6-1. Shops sell tube well pump heads in reassuring colors such as green, blue, and grey. (Photo credit: Suzanne Hanchett, 2006)

Researchers in the Matlab Subdistrict of Comilla District found that it was confusing to people when arsenic was described as a "poison." So they used more culturally appropriate ways to communicate with the public about the health risks posed by arsenic. One ethnographic researcher, Ms. Aasma Afroz Shathi, of ICDDR,B's Matlab Project, has observed that most people assume that "poisonous" materials (pesticides are the most familiar) have distinctive smells and colors. Such poisons kill quickly when ingested. It is, she says, a sort of miscommunication to use the word "poison" in connection with arsenic, because people have difficulty grasping the idea that a "poison" can have no color or special taste. It seems impossible to many people that good-tasting water could be "poisonous." In pursuing this matter with other professionals, we have found that most disagree with this point of view and think that it is appropriate to use the word "poison." Several said, however, that they do modify the description, perhaps by saying it is "one type of poison," rather than simply using the blanket term biish. In one group discussion an NGO worker expressed the view that it was quite difficult to translate the phrase "slow-acting poison" into Bengali, although that is the expression that is needed. (Hanchett 2006)

Staff of a Columbia University cohort study in Araihazar Subdistrict (Narayanganj District), told us that they avoid the word "poison" so as not to excessively frighten people. The project director in 2005-2006, Dr. Tariqul Islam, said, "We use the color sort of idea to explain about arsenic in water. We say it's like color, fertilizer or pesticides." Comparing arsenic to dye or color makes sense in this area, he claimed, since many of the people in the areas where they work are involved in the textile industry. The point came up that makers of pesticides actually refer to their product as "medicine" (*oshud*). He mentioned also that arsenic is an ingredient in some homeopathic medicines, so many would resent its being called a "poison." (Hanchett 2006:24)

Over the first decade of the 21st century, despite all of the miscommunication and institutional slippage, the affected public gradually has come to understand, however vaguely, the arsenic problem and its causes. Large gaps in awareness still exist. Over time we have come to hear many villagers speak of their water sources as either "arsenic-free" (*aarshonik-mukto*) or "arsenic-contaminated" (*aarshonik-jukto*), as we mentioned in Chapter 4.

This understanding has come in various ways to various people and their communities. Arsenic-related cancers are especially alarming and tend to alert whole villages to the seriousness of their drinking water problem. Solutions, however, remain elusive in many cases, as service levels are shockingly low or non-existent in most rural areas. Public frustration levels are high in arsenic-affected areas, because those who recognize the problem have great difficulty getting accurate information from government or other experts.

Popular Interpretations of Arsenic-related Skin Disorders

Arsenicosis sufferers have a difficult time being socially accepted in almost all the places we have visited. Even if they "know" about arsenic as a cause of illness, people may be reluctant to touch, take food, or share a bed with an arsenicosis patient having skin lesions, gangrene, or other visible and ominous symptoms. High percentages of people interviewed express reluctance to form marital connections with families of arsenic patients. (Hanchett 2006:12) There are practical, livelihood issues as well, as arsenicosis often leads to fatigue.

In a 2004 visit to an arsenic-affected village in West Bengal, we heard of a case in which a person with arsenic-related illness was ostracized even in death. Neighbors and everyone assumed that some sin or curse of God had caused the death. During the cremation, it was reported, no one came to touch the body.* Such strong social rejection, reported in many newspaper articles and case studies, can only be the result of cultural beliefs about the moral causes of arsenic-related illness, especially skin disease.

Such folk ideas about health are based on pervasive beliefs about the relation between bodily health and morality. Medical anthropologist Arthur Kleinman suggests that we view such views as "symbolic systems built out of meanings, values, behavioral norms, and the like." (Kleinman 1978:86) Folk health concepts of both Muslims and Hindus in Bengali speaking villages connect sin or other types of moral transgression and illness, although specific ideas about these connections differ from one person to another. Illness thus caused is generally assumed to be transmitted from generation to generation here, as it is elsewhere in South Asia.

Skin disease is traditionally viewed as indicating moral transgression or sin in the present or earlier generations of a lineage (*bongsho*). In Laksham Subdistrict, for example, almost all women interviewed, and many men as well, expressed the view that skin diseases result from sins (*paaper fal*), such as adultery, dishonesty, or suicide, and that this curse is transmitted from one generation to another (We also have been told that the affliction may skip one or more generations.).

One community health worker in Laksham told us that, "Non-patients think 'arsenic' is a disease inherited through the patrilineal kin group (*bongsho*)." Those who were identified as arsenicosis patients through the local project were taking practical steps to help themselves, he said, but others were indifferent. Unfortunately, we also found a large number of non-patients to be complacent about their drinking water. One female folk healer

*This report was from a village in Deganga Block, West Bengal, in June 2004. The information was provided by our guide, Mr. Priyotosh Mitra, then working as a consultant to UNICEF. (Suzanne Hanchett and Shireen Akhter notes)

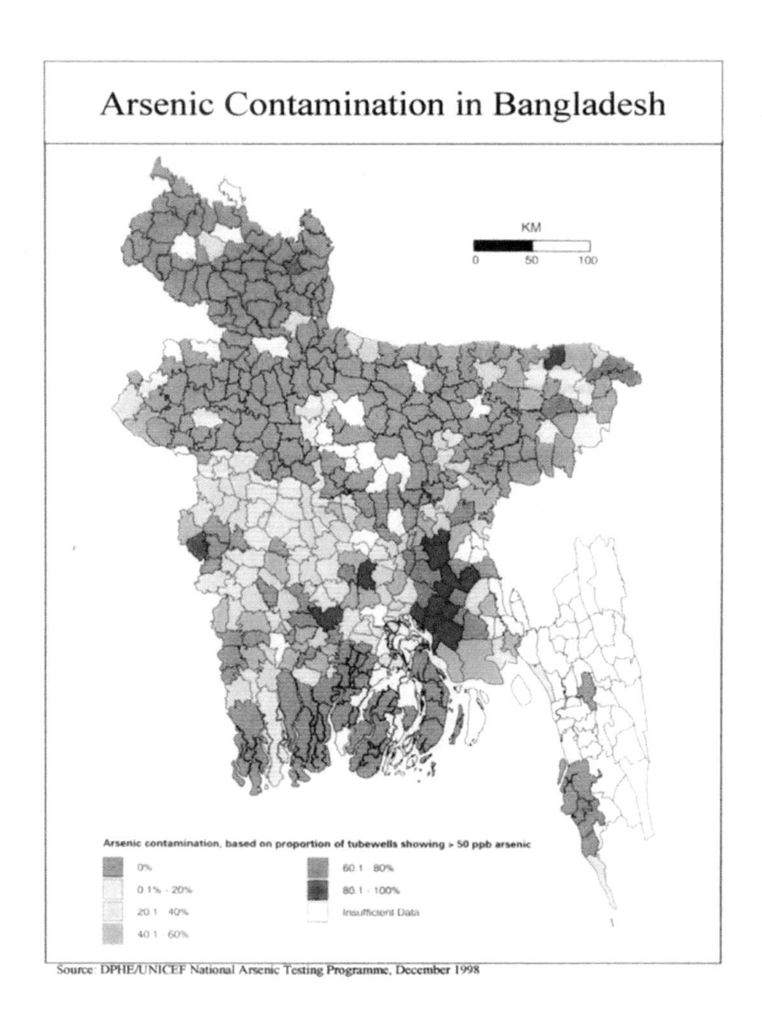

Map 6-1. Arsenic Levels in Bangladesh Ground Water (Government of Bangladesh, Department of Public Health Engineering, and UNICEF)

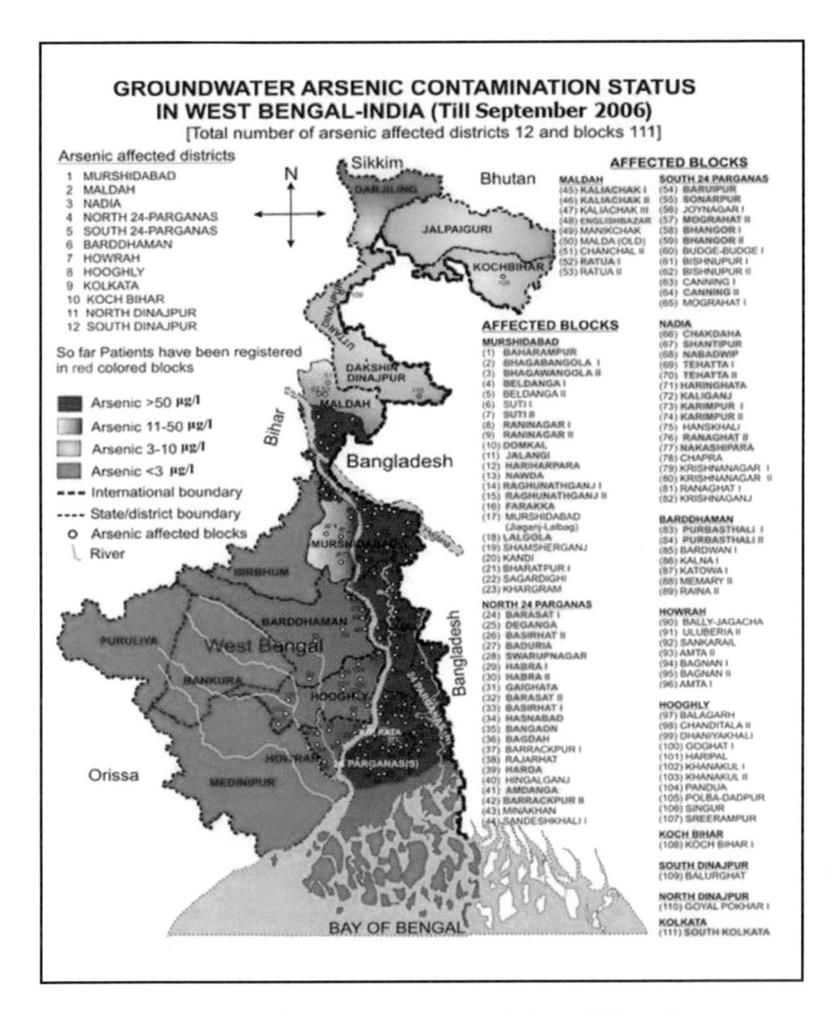

Map 6-2. Levels of Arsenic in West Bengal Ground Water (Source: Jadavpur University, School of Environmental Sciences)

(kobiraaj) said of her family in 2009, "We are not arsenic patients, so we can drink [arsenic-affected] tube well water."

In Bera beliefs associating hereditary sin and skin disease were found to be generally similar to those in Laksham, but they were not identical. There was some question about whether both white and black spots were caused by arsenic, or just black spots. One Bera explanation for white skin discoloration, which came up in two or three different interviews, had to do with refusing to share milk or yoghurt. We were told, for example, that "If someone gets skin disease, some fault-finder will say this happened because of sin (paaper fal). Some people believe this happens because of God's will. If there is some conflict between two people, and one has skin disease, the other will say, 'You misbehaved with me. That is why you got this skin disease, from your sin'."

Another 2009 comment from a Bera interview connected the color of skin spots with the moral cause. "Whiter spots on the skin are the result of sin....If you have some yoghurt or milk in the house, you must give it to your relatives when they visit. If someone asks you for that and you refuse to give it, or you say you don't have it, then you may be attacked by these skin diseases. People will say, 'She did not give milk/yoghurt, so she gets white skin disease'. It's a sin, not to give to others. If you do bad work you may be attacked by these diseases. People say *paaper jonno hoyse* (It is caused by sin). There is no medicine for these diseases."

These speculative comments show that some symptoms of arsenic poisoning, especially skin discoloration, lesions, and gangrene, were already explained by popular oral traditions and folk healers' principles of practice before they were connected to arsenic poisoning. Because skin disorders are among the most visible symptoms of arsenic poisoning, in 2009 we interviewed groups and key informants about skin and its disorders, and the causes and treatments for skin diseases. More than ten types of perceived skin disorders were identified in our 2009 interviews

Case Study 6-1. One Family's Mixed Practice

Photo 6-2. This rain water harvesting unit provides safe water to a household of 10 in Muradnagar Subdistrict for nine months of each year.

Photo 6-3. The mother of the family, however, cooks all meals with arsenic-contaminated water. She did not participate in village discussions of the arsenic problem. Her daughter-in-law (left) is aware of the problem. (Photo credits: Suzanne Hanchett, 2006)

in Laksham and Bera Subdistricts. Besides familial immorality, other frequently mentioned causes of rashes, pimples, and other skin disorders were dirty water, "heating" foods, seasonal change, a Hindu goddess (in the case of smallpox), and arsenic. We found that ideas about arsenic-related skin disease varied greatly from place to place. Some people, even in areas where arsenic mitigation projects operated, thought that arsenic caused skin problems only on the palms of the hands and soles of the feet. It does cause such problems, but other skin symptoms were disregarded. It is possible that misinformation from project staff or other professionals contributed to misunderstandings, because they too are not always as well informed as they should be.

These findings do not explain everyone's response to the arsenic problem, but they do explain many of the social problems that arsenicosis sufferers experience, especially social ostracism. One Bangladesh study found that,

> Far from being nurtured and pitied by their communities, sufferers from arsenic poisoning are shunned to the point of becoming social misfits or outcasts. The stigma is similar to that of leprosy, and our research indicates a condition that is threatening and confusing to all concerned, amounting to the "spoiled identities" outlined by Goffman (1968). (Hassan *et al.* 2005)

There are symbolic connections among some of the popular explanations for skin disease. Dirt and pollution is one common theme. We have been told on many occasions—in Pabna and Comilla Districts and also in southeastern Bangladesh—that pond water or flood water having urine, feces, or garbage in it will cause skin disease. Just as impure water is thought to cause skin disease, pure rain water can cure it, or at least cure the type called *ghaamaachi,* as discussed in Chapter 5. Moral transgression or sin is another theme. This is assumed to defile a person's soul and therefore also his or her body, indeed, even the bodies of future generations.

Concerns about Contagion

Concerns about whether arsenic related illness is contagious are found in several studies of people's responses. One early case study by Dr. Mahbuba Nasreen provides an example:

> *Kalimul Islam (age 40) is a victim of arsenicosis in Charlal Village. He is a jute mill worker with a monthly income of Taka 5000. He was very happy with his family even with his small income. Three years ago he was attacked by arsenicosis. At the early stage he was not serious about the disease because he thought it as skin disease. However, people in the village and his work place gradually avoided him because they thought Kalimul had a [contagious] disease. ...He is [losing] physical strength and becoming psychologically depressed because of the disease.* (Nasreen 2003:349)

Farhana Sultana presents the case of a woman whose marriage was destroyed by the arsenic problem:

> *Rashida was married at a young age and came to live with her husband in this village. She drank water from the tube well in the courtyard, as did the rest of the family. (A) Few years ago, Rashida started to show symptoms of arsenicosis, and continued to get worse, as keratosis and melanosis showed up all over her body. Fearing that she was contagious and cursed, her husband remarried and brought home a second wife. This wife also started to show similar symptoms of arsenicosis*

recently, and the tube well water was tested and found to contain high amounts of arsenic. Rashida's husband has now abandoned both wives, and taken a third wife and lives in the city. Rashida has no source of income except for the meager earnings of her eldest son; her other children are too young to work. Rashida spends most of her day unable to do much, in considerable pain, and relies on external charity and support for her medical expenses as well as household expenses. (Sultana 2006a)

A 2005 study by Farhana Sultana in four subdistricts found that income, exposure to media, and literacy influence opinion about whether or not arsenic-related illness is contagious. Poorer women and men are more likely to believe that it is contagious than more solvent people. (Sultana 2006a) Although the majority of Sultana's survey respondents expressed the view that arsenicosis is not contagious, "It would often transpire that even if people originally agreed that arsenicosis was not contagious, they sometimes expressed fear that it might become so and that they would fall ill if they socialized with an afflicted person." A "substantial minority" of respondents in all socio-economic categories expressed the view that arsenicosis patients should be isolated from society. Sultana found the following percentages of males and females answering Yes/No to the question, "Do you think arsenicosis is contagious?"

	Hard-core Pool		Poor		Lower Middle		Upper Middle		Rich	
	M	F	M	F	M	F	M	F	M	F
Y	6	17	12	23	15	11	0	0	0	8
N	94	83	88	77	85	89	100	100	100	92

n=232: 98 males (M), 134 females (F) (From Sultana 2006a, Table 2.4)

A study by Jan Willem Rosenboom reported a similar finding. (Rosenboom 2004) In Bangladesh average literacy and education levels are directly related to socioeconomic status, so less educated and illiterate people are more likely to view arsenicosis as contagious. In most socioeconomic categories women are more likely to hold this view than men.

Impacts on Women's Work and Women's Status

Sultana's research emphasizes the ways that reduced safe water access has affected women's work and women's status. One effect has been to increase women's water-carrying burdens. Although one-third of Sultana's 232 interviewees said that men occasionally helped with the task of fetching drinking water from distant sources, she found that the arsenic situation reduced men's willingness to help in the majority of cases. She quotes one man as saying something we also have heard: "I would die before I fetched water for a woman. If I did, people would think I am mad." (Sultana 2009:437)

Modesty concerns and purdah norms restrict women's movements outside of their immediate neighborhoods, limiting access to safe sources. One teenage girl is quoted as saying, "My father said we'll have to drink this water [from the red-painted tube well] and that we shouldn't go to the bazaar to get water from the green tube well. It is not allowed." "Such sensitivities," explains Sultana, "often result in entire families continuing to consume contaminated water in a trade-off between safeguarding family honor and taking the risk of consuming safe water." (Sultana 2009:432)

Case Study 6-2. Setting Aside Status Concerns to Get Arsenic-free Water

Sayed (fictitious name) *said that when tube wells were first marked to show their arsenic status, "We became very worried, because all the tube wells in our* baaRi *[homestead compound] were marked red. But no one explained what the red marks meant. I was not there when they did the tests. My relatives in the* baaRi *just said that it was now forbidden to use the water from those tube wells. Being uninformed, we perceived arsenic in our own way. I was thinking about what to do. I consulted some senior relatives and neighbors. I decided that boiling the water would probably destroy both germs and poison. So, for four months we boiled the water from our red-marked tube well. After that I stopped doing it. It is a very tedious and costly chore, and it adds to women's household burdens. There was a deep tube well in a neighboring* baaRi. *Our women normally did not go outside of our own* baaRi *to collect drinking water, but now they do."**

Case Studies 6-3a-b. Consuming Arsenic-contaminated Water to Protect Social Status

While our information is not consistent on this point, two situations we encountered show that socioeconomic class status concerns may at times interfere with access to safe water. Ironically, both of these situations resulted in high status people continuing to drink arsenic-contaminated water.

6-3a. *In one village, a wealthy family refused to take water from a deep tube well installed near to—but not inside—the compound of another wealthy family. This was despite the fact that the deep well had arsenic free water and their own homestead's*

*A 2006 report from Ramganj Subdistrict, Lakshmipur District, by Tofazzel Hossain Monju

well was not safe. According to women of the family, they had no quarrels or other problems with the other family, but their status would have been demeaned by taking drinking water daily from the neighbors' deep well.

6-3b. In the other situation we heard strong complaints from a group of high status union council members in Laksmipur District that the Danish aid agency, Danida, had installed arsenic-safe deep tube wells only in the compounds of poor people. We were told that this Danida policy discriminated against rich people and caused what they called "social problems." *Because "our egos absolutely will not allow us to ask for water from the compounds of poor people," the council members explained, "we are forced to drink unsafe water." Expressing an astonishing (to us) sense of social entitlement and indifference to the special needs of poor people, they argued that it was unfair of Danida not to help them, because "everyone needs safe water."* (from Hanchett 2012)

Sultana observed "a few instances where a safe tube well was in the homestead of a poor family," when some wealthier women "were forced to overlook such social status infractions to have to depend on the poor in an odd reversal of power relations." Others were reluctant to take water from people of lower status, as this "went against the sensitivities of most of the wealthier households." (Sultana 2009:434-435)

We have encountered both types of cases, but the latter seems to be more common at this time. These findings suggest a need to address directly people's status concerns when educating the public about the arsenic problem and ways to get safe water.

Technical Solutions

Four different types of technical solutions are recognized: 1) changing to arsenic-safe groundwater; 2) consuming arsenic-free surface water which has been treated to remove pathogens; 3) rain water harvesting; and 4) removal of arsenic by some form of filtration-*cum*-chemical treatment. Each of these poses challenges that are technical, social, and/or administrative.*

Changing to arsenic-safe groundwater is done in two ways. One is to return to the formerly popular dug well. Unless they are tightly covered, however, dug wells have been found to become contaminated with bacteria over time. At least one West Bengal project, nonetheless, is continuing with this approach.** The officially preferred method in Bangladesh is to install deep tube wells, if they are technically feasible. Unlike the more widespread tube wells that draw water from shallower aquifers, they are very expensive to drill and thus cannot be installed in every compound.

Social relationships affect who has access to deep tube wells. It is considered most practical to drill them near to or inside residential compounds so that specific families will take responsibility for maintaining them, but social relationships limit access if they are in sections of the homestead that are considered private. The official promotion of deep tube wells has interfered to some extent with testing of alternatives, such as pond sand filters or rain water harvesting units. According to one BRAC study, "When villagers got deep tube wells they started losing interest in other options." (Jakariya 2003:33)

Surface water, the water of ponds or rivers, can be filtered and/or treated chemically to make it safe to drink. For a while the pond sand filter (PSF) was widely promoted as a technical solution likely to provide a convenient and safe drinking water alternative to the tube well. We have visited and assessed approximately 50 pond sand filters in arsenic-affected areas of Bangladesh. They worked technically, but a large majority did not continue in use for more than a couple of years at most. The main reason for

*One experimental project is testing a method to remove arsenic from aquifers, offering the possibility of a fifth type of safe water option (Dr. Richard Johnston, personal communication)

**Project Well

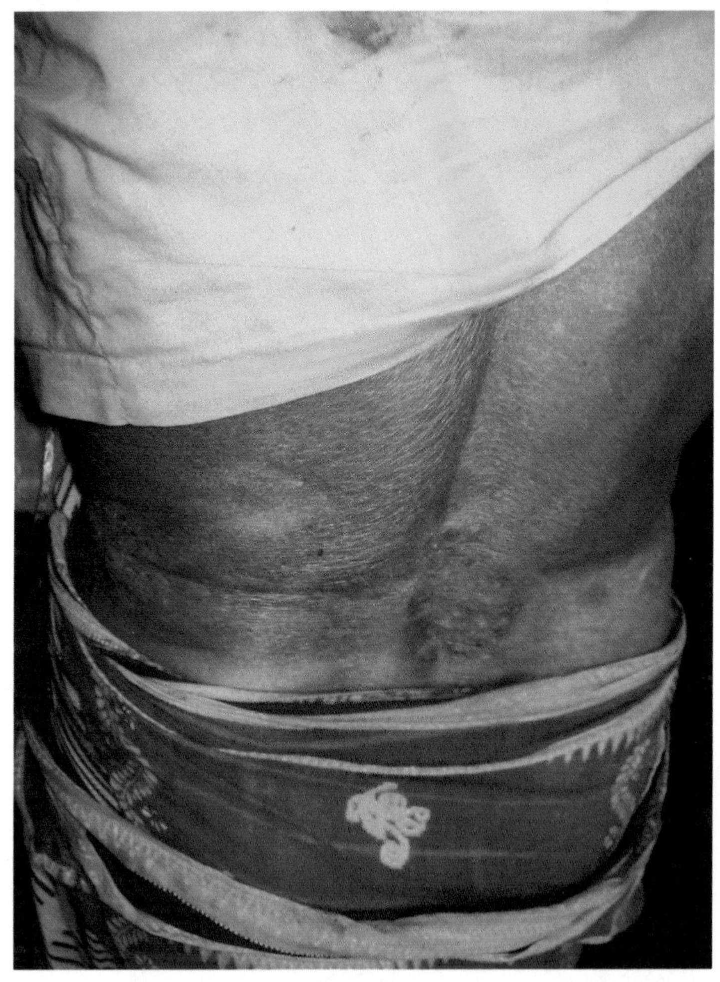

Photo 6-4. Darkened skin area, a symptom of arsenic poisoning (Photo credit: Anwar Islam)

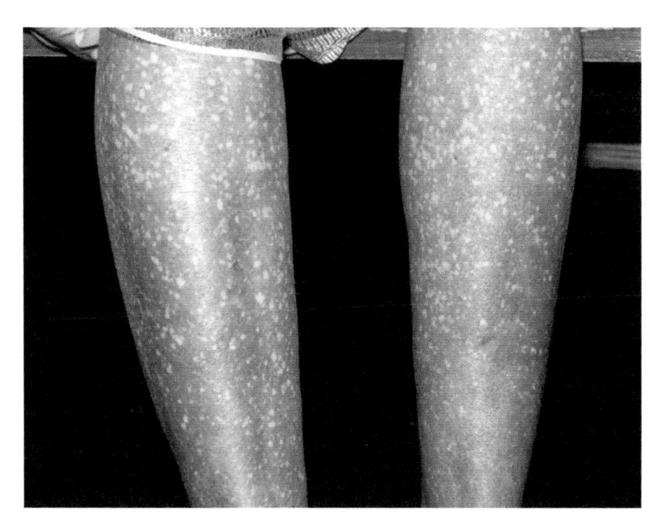

Photo 6-5. Light spots on the skin related to chronic arsenic poisoning (Photo credit: Kazi Rozana Akhter)

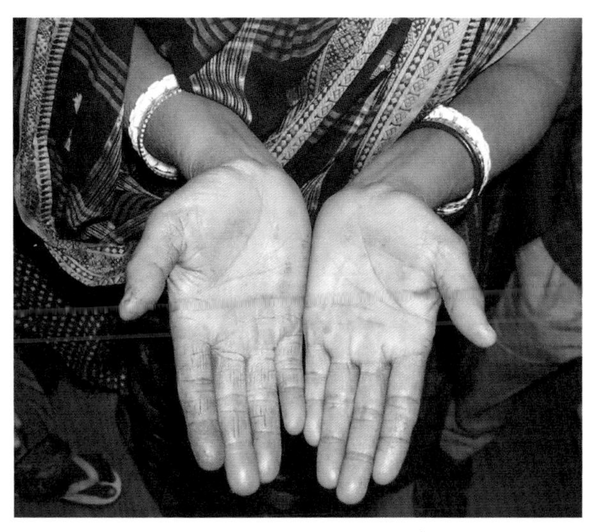

Photo 6-6. Hardened spots on the palms of the hand, called keratoses, are one symptom of chronic arsenic poisoning. (Photo credit: Suzanne Hanchett)

Photo 6-7. Lesions on the feet make walking painful for arsenicosis patients. (Photo credit:Anwar Islam)

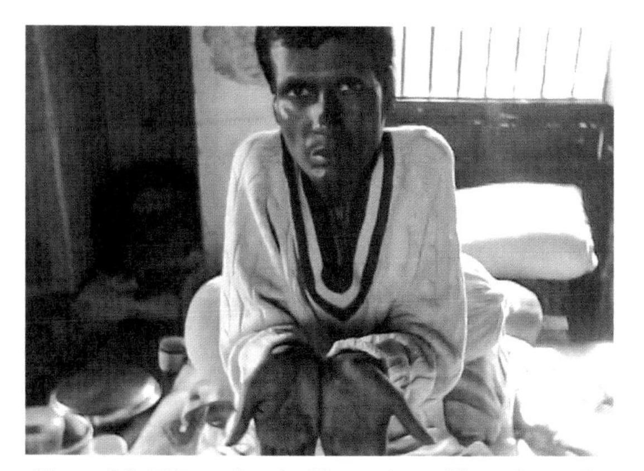

Photo 6-8. This seriously ill man is a village doctor in Muradnagar Subdistrict. One symptom is keratosis on his hands.

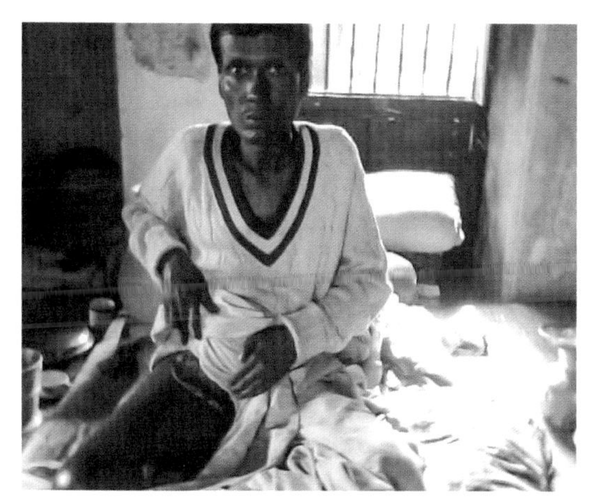

Photo 6-9. Cancerous lesion associated with arsenic poisoning (Photo credits, 6-8, 6-9: Pathways Consulting Services Ltd., 2006)

failure was the opportunity cost of giving up profitable fish culture. The social context made it almost impossible to solve this problem. Most ponds are jointly owned by groups of siblings or cousins, any one of whom has the right to use the water body for commercial fish culture, which is not compatible with pond sand filtration because the required chicken manure makes the water unpotable even with filtration.

Another reason for PSF failure is the availability of alternative water sources. A common situation is government-funded installation of a deep tube well in the vicinity of a pond sand filter. Tube wells require less maintenance and less community cooperation than pond sand filters, and they are therefore more desirable in general. In one Muradnagar Subdistrict village we visited twice, we found that villagers lost enthusiasm for their pond sand filter when household-level arsenic removal filters were introduced, even though the household filters were not affordable to all.* (Photo 6-14) We have, nonetheless, seen a few cases in which pond sand filters did work quite well. Two surviving pond sand filter systems we found were donated for general village use by a pond's single owner. One filter established by UNICEF at a newly purchased pond also appears to be succeeding.

Other technical solutions include rain water harvesting and several types of filters. Filters treat water either inside the home or in a plant used by a whole neighborhood. Filtration of arsenic-contaminated groundwater is not a simple process, as arsenic salts dissolve very thoroughly and require chemical treatment with media that are too expensive for poor people.

---

*We visited this village once in 2006 and again in 2009. During the first visit a women's committee spoke proudly of their achievement in getting and maintaining their pond sand filter. Three years later the most well-to-do women had gotten household filters. It is likely that their families had a financial interest in fish culture, but we did not investigate this case deeply. Some of the poorer women on the non-functioning committee were unhappy at the loss of access to safe water.

The Government of Bangladesh has tested and certified at least seven types of arsenic removal filters as sufficiently effective for sale to the general public. Thousands of arsenic-removal filters have been distributed for free or at subsidized prices to households and communities on a trial basis by governmental programs, UNICEF, research organizations, and NGOs. One West Bengal project installs and maintains arsenic removal units that filter water from publicly accessible tube wells.*

Our team in 2009 conducted an assessment of the social acceptability of four types of arsenic-removal filters, three for use in households and the Sidko plant, a neighborhood installation. Water from the filters, which had been provided at highly subsidized prices, was the main household drinking water source in 70 to 85 percent of surveyed households. Pond water, however, continued to be popular for cooking. In one village where arsenic-removal filters had been distributed to households with arsenicosis patients, however, we found that the treated water was drunk mostly by people with symptoms, while others continued to drink from arsenic-contaminated sources.**

The short-term, project-based approach of service providers continues to be an obstacle to problem solving. In one village, for example, local people had been trained to test water for arsenic content and supplied with test kits. We met one young man who had been trained and then set up a Sidko plant outside his home. He was carefully maintaining it and monitoring the arsenic content of the water on a regular basis. But when the project ended, Department of Public Health Engineering officers took away all water testing equipment and supplies, including his, and stored them in a warehouse. When asked why they took away these

*Bengal College of Engineering, Project to Remove Arsenic from Village Drinking Water Supplies, Prof. Anirban Gupta, Director, interviewed in June 2004

**Findings are reported in Planning Alternatives for Change and Pathways Consulting Services Ltd. 2009, a report submitted to UNICEF Bangladesh.

important water test kits, an official said they were government property and needed to be returned after the project was finished.

In rural areas we have seen many abandoned, broken structures—pond sand filters, rain water harvesting systems, and others. These were once technically sound, expensive objects that have failed because they were not socially viable, or because of inadequate follow-up, monitoring and maintenance. Some fell into disuse when a small part broke. New shallow tube wells are being sunk in high risk areas without groundwater being tested for arsenic. Many people (no precise count is available) who shifted to arsenic-free sources at first have returned to their earlier arrangements.

The Way Forward

A somewhat positive national survey finding from 2009 was that, in 13 randomly sampled Bangladesh districts with arsenic problems, consumption of officially "safe" water ranged from 52 to 84 percent, with a median of 71.5 percent.* This survey combined rural and urban areas. Although there are no comparable figures from earlier surveys, these findings and other studies demonstrate that some families, at least, are finding ways to drink arsenic-safe water.**

People get their ideas about their water through social networks, from project field staff, from mass media, and from their own deliberations. However much they know, people's

*Data from the UNICEF Bangladesh's Multi-Cluster Indicator Survey on the following districts (with percentages of households consuming water with an arsenic test value of less than 50 milligrams per liter): Brahmanbaria (69.8), Chandpur (56.3), Comilla (51.5), Feni (71.5), Lakshmipur (77.3), Noakhali (60.5), Madaripur (68.3), Manikganj (79.3), Chuadanga (71.9), Jessore (73.2), Meherpur (83.7), Satkhira (72.5), and Sunamganj (52.1). Source: Progotir Pathey, 2009.

***Progotir Pathey* 2009, Table 19

capacity to respond to warnings about arsenic is constrained by their situation: their social status, education, money, and women's time. Social considerations thus strongly affect the public's response to arsenic mitigation efforts. And existing cultural views about arsenic-related illness—which have deep roots in folk thinking about the body and the family—deserve further attention by concerned development professionals.

Poor people's special circumstances also deserve careful consideration. They are known to be particularly vulnerable to arsenic-related illness and thus are at especially high risk of arsenic-related death. They tend to have much less education than others and thus generally less access to information about arsenic. Their ability to arrange good quality water tends to be weak, so they depend on fewer sources. They need community-managed water points because their economic position makes private supplies unaffordable. And men whose livelihoods depend on manual labor need to drink large quantities of water.

There is a need for further research on some key points, especially (1) whether and why people do or do not stop using arsenic affected water sources when alternative options become available and (2) the experiences of girls in arsenic affected areas, specifically whether girls are as adequately covered as boys by health screening and treatment programs.

Considering that most alternative water sources are used by groups, not individual households, community level organizing, if it is done at all, is too weak, at least in many Bangladesh localities. This will affect the long-term viability of arsenic mitigation systems by jeopardizing maintenance arrangements. We have not investigated community organizing in West Bengal arsenic mitigation projects, but this topic also deserves close attention.

Since women are the primary managers of domestic water in rural areas, any effective local-level program must include women as active participants in planning alternative water source placements and characteristics.

Service providers and policy makers themselves share a large part of the responsibility for the inadequate state of arsenic mitigation, yet professionals tend to blame the public's lack of awareness or motivation. Poor planning and coordination is one of the most challenging problems. Numerous types of agencies—governmental, UN, NGO, religious, and volunteer groups—have rushed into villages to implement schemes in an uncoordinated manner. Sometimes two or more organizations have offered competing or conflicting services and messages in one place. Their differing messages, tube well testing methods and results, and ideas about how to solve the problem all too often confuse the people they intend to help. In Bangladesh, there has been some progress in reducing such uncoordinated approaches, as required by the National Policy (Government of Bangladesh, 2004), but numerous problems still remain. Different Bangladesh ministries (Health, Environment, Local Government, and so on) do not share arsenic-related information or coordinate their services to a sufficient extent.

Well-managed arsenic mitigation services—rather than projects—should be based on social and cultural insights, not just on technical expertise. Such services would have the potential to influence people's attitudes and understandings, even if they did not entirely change them. Awareness alone, however, is not always enough to motivate people to change their water use habits. Status concerns and community social organization strongly influence water sharing arrangements, and these need to be part of any local mitigation planning. For example, solving the arsenic problem may not be a family's priority, especially among poorer households. Most challenging are the cultural and emotional issues associated with beliefs in contagion, the fear of supernatural curses, and the

Photo 6-10. An improved type of dug well, protected from contamination but open to air. Water is collected by means of a hand pump.

Photo 6-11. Another type of improved dug well (Photo credit: Anwar Islam)

Photo 6-12. A "ring well" is a fully covered dug well with water accessed by a hand pump. Bera, 2009 (Photo credit: Akash)

Photo 6-13. Woman drawing water from a pond sand filter (Photo credit: Suzanne Hanchett)

Photo 6-14. A member of a pond sand filter com-mittee showing us the village's informational poster in Muradnagar Subdistrict in 2006. Three years later, the committee was no longer functioning, and the filter was not in use. (Photo credit: Suzanne Hanchett)

Photo 6-15. A household arsenic removal filter produced in Bangladesh, the Sono Filter, which in 2009 cost around 2500 Bangladesh taka (about US$40), needs replacement every couple of years. (Photo credit: Kazi Rozana Akhter)

Photo 6-16. A small arsenic removal filter is the READ-F, which in 2009 cost around 5000 Bangladesh taka (about US$85). The filtration medium needs replacement at a cost that was almost the same as the original purchase price.

Photo 6-17. The Sidko brand community-level filter can provide arsenic-free water to as least 100 households. Its success depends on a location that is accessible, the financial ability of the community to replace periodically the filtration medium, and a local maintenance routine. In 2009 media replacement cost Bangladesh taka 18-25,000, about US$300-400. (Photo credit: Suzanne Hanchett)

Photo 6-18. A village woman performing routine maintenance on a Sidko arsenic removal filter. Bera Subdistrict, 2009 (Photo credit: Shireen Akhter)

Photo 6-19. This Muradnagar Subdistrict village water tank is connected to a pond sand filter and formerly distributed piped supply water to 15 homes. When a pipe from the pond to the tank broke, however, it was not repaired; so the expensive system is not in use.
(Photo credit: Suzanne Hanchett, 2009)

Photo 6-20. This expensive and well-built in-ground rain water harvesting unit was never used. The reasons are unclear. One person of the locality said someone had urinated on top of the concrete cover. (Photo credit: Pathways Consulting Services)

widespread assumption that one unhealthy individual brings dishonor or bad luck to all his or her relatives.

An arsenic mitigation program, project, or service is a social change effort at least as much as it is a technical exercise. Communication and social difficulties can be overcome through careful, participatory planning, staff training, and management. The people who implement mitigation activities have a responsibility to increase their own awareness of the social causes and consequences of the arsenic problem and to incorporate this awareness into their approaches.

Summary

This chapter reviews the issue of arsenic in drinking water in Bangladesh and attempts to mitigate the problem since the late 1990s. Culturally-based ideas about the body and the ominous meanings of skin disorders strongly affect the social response to people suffering from arsenic-related illnesses. Skin disorders of some types, for example, are generally assumed to be the result of ancestors' sins. Strengths and weaknesses of technical solutions are reviewed: changing to arsenic-safe groundwater, consuming treated surface water, rain water harvesting, and removal of arsenic by filtration and chemical treatment. Common perceptions of arsenic have changed somewhat over the past 10-15 years, but the weakness of the government's and donors' responses to the problem has left many affected locations without enough in-formation or help. Too much mitigation work has run on the short-term "project" model, and there are not enough ongoing services.

7. Water's Powers: A Schematic Overview

There is considerable wisdom in folk beliefs despite their lack of connection to scientific traditions. Water—especially water contained in rivers or ponds, tanks, or lakes—is treated and spoken of as if it is alive and has a personality. It gives and takes. It can be insulted. It can be respected and adored. It can do good or harm.*

The preceding discussion has presented many people's statements about their water and our observations about the social and cultural context that affects domestic supply water access, sharing, and uses. There are clear themes in these comments, but each family organizes its water life in its own way, so our information is contradictory at times. The purpose of this chapter is to highlight some patterns in the complex and varied beliefs and practices described in this book.

Some meanings encoded in the symbolism of Bangladesh and other South Asian water practices and lore are summarized below in Table 7-1.

*Some of these ideas and themes can be found in South Asian literature. In one short story by a Bengali author, for example, water can be "killed" by polluting it with toxins (Mahmud Rahman 2010).

Table 7-1. Water's Powers, Qualities, and Meanings

Water's Power, Quality, or Meaning	***Illustration: Cultural Practice, Folk Belief, or Story***
Capacity to form a personality or become a living entity	•Helpful or harmful actions by rivers, ponds, tanks, or lakes •Can take on variable "hot" or "cold" properties, as humans do •Rivers have gender (mostly female, but some male)
River as deity	•Hindu belief: Ganges River identified with the goddess Gaŋgaa
Transformative Can produce changes in status Divides one status/condition from another	•Bathing/Washing: removes "pollution," creates a state of "purity" •Washing/pouring water in rites of passage, such as birth, marriage, death
Carries away sins Absorbs and removes sins from a person's body and soul	•Hindu and Muslim pilgrimages to bathe in special waters •*Jamjam/Jumjum* water used to purify household pond water •Hindu practice of carrying water from pilgrimage sites

Water's Power, Quality, or Meaning	***Illustration: Cultural Practice, Folk Belief, or Story***
Absorptive Capable of absorbing spiritual messages, thoughts, powers, holiness, and personal flaws	•Water from Mecca carries holiness •Putting an an amulet (*taabiij*) which contains a written prayer into water, which then gains healing power •*paani paRaa*: blowing or speaking a prayer over some water, which then acquires the power to bless or heal
Healing, protective, auspicious	•*Jamjam/Jumjum* water from Mecca: -Used to heal illnesses -Sprinkled in house to protect it from evil spirits or possession by dangerous ghost powers •Healing waters of special ponds •Ponds with power to give fertility
Enforces moral codes A demander, provider, and challenger of humans A home for spirits who give to and take from humans	•Large Ponds (*dighi*): -Produce wealth -Respond to misdeeds, especially violation of moral codes (Kanchanpur

Water's Power, Quality, or Meaning	*Illustration: Cultural Practice, Folk Belief, or Story*
	Dorgabari and other large ponds produced vessels until someone stole a spoon or made some other mistake) -Require human sacrifice before they will be filled with water (two examples in Chapter 3)
Power increases as volume increases	•Comments that a large volume of water in ponds, lakes, and rivers ensures "purity"
Strong flow improves quality	•Comments about water being especially good and "new" during the rainy season
Dynamic, moving force Metaphor for time, tradition, flowing mother's milk Change-*cum*-continuity in a family	•River imagery in proverb: "A black daughter from a good family is good; and muddy water from a [big, strong, high volume] river also is good." •Jamalpur District reported custom of burying placenta near a pond to ensure adequate flow of mother's milk (Blanchet 1984)

8. Conclusions

"A crucial step in laying the foundations for sustainability...is deepening individual and collective understandings of how water is understood and valued." (Johnston 2012:xvi)

When we speak of water as "a common resource for the sustenance of all," as Vandana Shiva (2004) does, we mean social and spiritual sustenance, not just physical survival. The strong and pervasive ideas about water in South Asia have deep historical roots in this old civilization. Several historical influences feed a lively and creative folk culture. Culturally based named categories, concepts, and practices of the sort described in this book are at the heart of village social, emotional, and spiritual life. Their meanings, unlike scientific definitions, extend outward, creating visions of connectedness between people and values, social reproduction and places. Every part of the South Asian subcontinent has its own variations on these themes, but people in all areas share the view that water is far more than a physical or chemical substance.

We have reviewed many of the ways that Bengali-speaking villagers endow water with special meanings and values. We have glimpsed at their special water vocabulary, water's role at times of birth, marriage, and death, and some folk ideas about water's healing powers. Women's and men's responsibilities for domestic water collection, storage, and use have been reviewed. As in other world regions, women tend to have the greater share of responsibility. It is mostly women who decide which water sources are to be used for drinking, cooking, bathing and other domestic purposes.

A scientist hoping to help solve problems related to domestic water supply needs to understand the cultural context and the perceptions and practices that are based on it. And he or she needs to engage women in all stages of the problem-solving process. Two benefits of this approach will be (1) appropriate service design and (2) effective communication. Our Case Study No. 6-1 shows, for example, that not involving women, especially senior women, in planning local services can result in the arsenic problem being ignored or inadequately addressed. In this case, the principal cook for a household of 10 continued to prepare meals with arsenic contaminated water despite the fact that some of her family members were aware of the problem. Communication about arsenic at first stressed its poisonous quality. But the common perception of a poison (Bengali *biish*) is that it kills someone right away, so the use of this word in public information campaigns probably was a mistake. Misunderstandings also arose in connection with household arsenic removal filters, whose water was considered by many to be too "cold"—in the humoral folk medical sense – for sick children to drink. So the arsenic-safe water was not used when children were sick.

An access-related problem we have discussed results from a reduction in the number of safe and suitable water supplies. In areas with high salinity (some human caused, some natural) this is a serious issue, as it is in areas where arsenic is found in groundwater. The tube well at first seemed to solve access problems in Bangladesh and neighboring areas of India, because it offered a relatively inexpensive way to avoid sharing the homestead courtyard pump with people of other households. The tube well thus eased social tensions for thousands of housewives and continues to be very popular partly for this reason (and partly also because it reduced diarrheal disease). The number of wells officially declared to be safe, however, shrank dramatically with the discovery of the arsenic problem. Reduced numbers of safe wells once again create social tensions and ultimately put non-owners at high risk of losing access altogether, unless they are fortunate enough to live in the few rural communities with piped supply connections.

Culturally-based beliefs and practices are of great importance in programs or projects relating to domestic water supply. This is true for domestic supply more than for large-scale water resources management, where respect for cultural diversity and the right to water usually is sufficient.

Rights declarations and commons-view thinking are used in the United Nations and by activists to support the incorporation of indigenous and folk beliefs and practices into local water planning.[*] Such an effort can strengthen grassroots control over this essential resource by valuing and respecting local knowledge and practices. In a few documented cases, people have been able to overcome the formidable political and economic obstacles to achieve this ideal.[**]

There are, of course, many practical and social, political and legal considerations influencing water access. Formal and informal sharing arrangements are based on principles of family, friendship, charity and social or political institutions. Well-recognized economic and political factors matter greatly. These issues are thoroughly discussed by many other writers.[***]

---

*UNESCO's 1982 Mexico City Declaration on Cultural Policies defines culture as, "The whole complex of distinctive spiritual, intellectual and emotional features that characterise a society or social group. It includes not only the arts and letters, but also modes of life, the fundamental rights of the human being, value systems, traditions and beliefs." (UNESCO 1982, Article 10)

**Some cases from India are discussed in Bywater 2012.

***One good example is the group of contributors to Lahiri-Dutt and Wasson's collection, *Water First*.

Governmental institutions and donor agencies have substantial responsibility to ensure the human right to water. As Catarina de Albuquerque, the UN's Special Rapporteur on the Human Right to Safe Drinking Water and Sanitation, has said, "Institutions involved in the water and sanitation sectors must be responsive and accountable for their actions, and decisions must be participatory and transparent. All groups and individuals concerned and all relevant stakeholders must be provided with genuine opportunities to meaningfully participate and must be empowered in these processes." (UN General Assembly 2013, para. 23)

Mismanagement of the arsenic problem at all levels of society, government, and the aid system poses a threat to access in affected areas. Making arsenic mitigation a "project" instead of an ongoing governmental program or service was a mistake. The WASH Sustainability Charter argues for long-term service delivery as an essential need. Their general declaration unfortunately sums up the situation in many arsenic affected areas: "Many of those who may have benefited in the short-term from WASH projects now have systems that are not working adequately, or have failed completely."*

Developing a system of proper services to cope with the arsenic problem poses daunting challenges. International donor funds have been largely exhausted. Normal tax revenues are needed to support information dissemination, water testing, monitoring of filters, and routine screening of patients with symptoms of arsenic poisoning. In Bangladesh, institutions such as the Department of Public Health Engineering, the Local Government Engineering Department, and the Ministry of Health would all have to participate and coordinate. Union councils and other elected officials would have to take responsibility for important local coordination and monitoring functions.

---

*WASH Sustainability Charter (available from http://waterservicesthatlast.org/news/new_wash_sustainbility_assessment_tool)

Even if political will could be mustered to overcome the obstacles to funding and coordination of services, social and scientific obstacles would remain. Assuming that a development program emerges that strives to ensure rights and access by respecting folk views, social and intellectual problems will arise to challenge the effort. The principal social problems relate to gender and social class. Women must be involved at all levels of decision making.

If there were a scientific argument in favor of respecting indigenous views, it might be more likely to occur. But once again obstacles arise. Humoral theories of "hot" and "cold" and concerns about spiritual "purity" and "pollution" have no place in contemporary scientific theory. Nor do ideas of skin disorders being caused by ancestral sins. The intellectual tendency of a well-meaning, educated development agent is to combat such beliefs, not to respect or accept them; but indigenous or folk philosophers, for their part, tend to trust their own, time honored ideas and practices. The paramcters, categories, and concepts of scientific and indigenous/folk thinking are almost entirely different.

The science supporting respect for indigenous views is more about the development process itself than physics, chemistry, or biology. As many development workers can attest, respectful communication fosters trust. A superior or domineering attitude fosters fear and inhibits two-way communication. Important local concerns may not be expressed or communicated to official decision-makers and government officers. In the case of arsenic removal technologies being suspected of producing "cold" water harmful to children's health, either such fears were not communicated to field level workers, or they were not taken seriously enough at first. Knowing of this concern, however, is helping development agents to find ways to overcome it and monitor responses to public messages.

Whether their views are respected by development agents or not, people will continue trying to cope with their environments and meet their water supply needs. Cultural principles are among their basic coping tools. These strong, basic assumptions are expressed both in words and symbolically. Change does occur, but it happens only when people's understandings and motivations change. Trusted cultural traditions (vocabulary, concepts, and practices) will not be erased by any short-term visitors, no matter how well-meaning. The best thing, then, is to respect differences and help to build on what is there.

The environmentalist movement has faced some of these communication challenges for a while. As strong as respect for water is, popular thinking will not always support scientifically-based environmental activism. The culturally based belief in water's strong purificatory power and its "motherly" qualities actually may create an obstacle in the case of the Ganges River, as Kelly Alley's research (2000, 2002) has demonstrated. Ironically, the strengths of the river are assumed by many experts to be so great, that it is thought not to need human help.

As water resources decline with global warming, there will be more and more pressure on limited waters. Solving this problem will require new approaches and new types of collaboration. Indigenous culture has supported adaptation to the environment, and it will be part of adaptation to change as well.

These are some of the reasons why to understand indigenous views. The material of this book demonstrates briefly how to do this. The approach must be broader than a typical public health or engineering one normally is. In folk thinking, perceived connections between social, spiritual and emotional life and the environment stretch in directions inconceivable to the scientific mind. But these connections give meaning to human action that science cannot provide. Language must be part of the learning and collaboration process. Ideas about the human body and health need to be acknowledged. Myths, legends and proverbs also get involved. We have applied this approach to water, but it can be used to explore any of the ways that humans connect to the plants, animals, and other features of their natural and built landscapes.

Indigenous and folk beliefs will be part of any effective solutions to the urgent water problems facing humanity. A scientist who is alert to folk beliefs and works in a participatory manner with the people directly affected by his or her projects has a good chance of promoting positive, sustainable change.

Glossary

Vocabulary related directly to water can be found in Appendix 2.

Bengali Term	Translation
aapabitro	"Polluted/Impure/Impurity" (Hindu word)
aaporishkaar	"Dirty/Dirt"
baaRi	A residential compound or homestead with multiple related households
baataash	"Air or "Wind" regarded as a physiological process related to health and illness in humoral medicine
bhijaa	"Wet"
bhut	A dangerous spirit or ghost
daai	"Midwife"
fokir	A religious specialist with healing powers and philosophical insights
ghaaT	Steps or platform at the edge of a river or water body
gorom	"Hot/Warm" (physical or metaphorical quality)
jinn	A dangerous spirit
kaãchaa	Anything raw or uncooked and unprocessed, including fruits, vegetables, and liquids. This term is also used to describe the vulnerable and unprotected state of a new mother.
kobiraaj	A folk healer, male or female
Kalima	Muslim verses from the Holy Quran and Hadith, compiled to instruct people in the basic precepts of Islam and widely read and memorized throughout the Muslim world

Bengali Term	**Translation**
naapaak	"Impure/Impurity/Polluted" (Muslim word)
oju	Ritual purification before Muslim prayer
porishkaar	"Cleanliness/Clean"
pabitro	"Pure/Purity" (Hindu word)
paak	"Pure/Purity" (Muslim word)
paani paRaa	A practice of folk healers or religious specialists: blowing on a container of water or reciting blessings over the water, on the assumption that the water will absorb the healing or blessing power of the words. The literal meaning of the Bengali phrase is "reading water."
paaRaa	Neighborhood, section of a village
pir	Muslim saint
pujaa	Hindu ceremonial offering to deities, a form of worship
*sochi (*or) *chochi*	"Polluted" (alternative, regional terms)
ThaaNDaa	"Cold" (physical or metaphorical quality)

Bibliography

Acharya, Amitangshu

2012 Managing 'Water Traditions' in Uttarakhand India: Lessons Learned and Steps Towards the Future. *In* Water, Cultural Diversity, and Global Environmental Change; Emerging Trends, Sustainable Futures? Barbara Rose Johnston, Editor-in-Chief. 411-432. Dordrecht: Springer, and Paris: UNESCO.

Afsana, Kaosar

2005 Disciplining Birth; Power, Knowledge and Childbirth Practices in Bangladesh. Dhaka: The University Press Limited.

Ahmed, M. Feroze

2003 Arsenic Contamination: Regional and Global Scenario. *In* Arsenic Contamination: Bangladesh Perspective. M. Feroze Ahmed, ed. pp. 1-20. Dhaka: ITN-Bangladesh, Centre for Water Supply and Waste Management, Bangladesh University of Engineering and Technology

Ahmed, M. Feroze, M. Ashraf Ali, and Zafar Adeel, eds.

2003 Fate of Arsenic in the Environment. Dhaka: Bangladesh University of Engineering and Technology, and Tokyo: The United Nations University.

Ahmed, M. Feroze, S.A.J. Samsuddin, S.G. Mahmud, H. Rashid, D. Deer, and G. Howard

2005 Risk Assessment of Arsenic Mitigation Options;

Final Report. Dhaka: Bangladesh International Training Network Centre (ITN) and Arsenic Policy Support Unit, MLGRD,C, Government of Bangladesh

Akhter, Kazi Rozana, and Suzanne Hanchett

2012 Managing Domestic Water in Bangladesh after the Arsenic Crisis. *In* Water, Cultural Diversity, and Global Environmental Change; Emerging Trends, Sustainable Futures? Barbara Rose Johnston, Editor-in-Chief. p. 249 Paris, Dordrecht, Heidelberg, London, New York: UNESCO-IHP and Springer. pp. 241-244.

Akhter, Shireen

2012 The Purity Problem and Access to Water in Bangladesh. *In* Water, Cultural Diversity, and Global Environmental Change; Emerging Trends, Sustainable Futures? Barbara Rose Johnston, Editor-in-Chief. p. 249 Paris, Dordrecht, Heidelberg, London, New York: UNESCO-IHP and Springer. pp. 246-247.

Alley, Kelly D.

2000 Separate Domains: Hinduism, Politics, and Environmental Pollution. *In* Hinduism and Ecology; The Intersection of Earth, Sky, and Water. Christopher Key Chapple and Mary Evelyn Tucker, eds. pp. 355-387. Cambridge, MA: Harvard University Press, for the Center for the Study of World Religions, Harvard Divinity School.

2002 On the Banks of the Gangā; When Wastewater Meets a Sacred River. Ann Arbor: The University of Michigan Press

Arthington, Angela H.
2012 Environmental Flows. *In* Water, Cultural Diversity, and Global Environmental Change; Emerging Trends, Sustainable Futures? Barbara Rose Johnston, Editor-in-Chief. 458-461. Dordrecht: Springer, and Paris: UNESCO.

Arthington, Angela H., R.J. Naiman, M.E. McClain, and C. Nilsson
2010 Preserving the Biodiversity and Ecological Services of Rivers: New Challenges and Research Opportunities. Freshwater Biology 55:1-16.

Asian Development Bank
2003 Current Status and Trends in Water Quality and Health and Social Impacts and Government's Policy and Institutional Frameworks, Strategies, Plans and Programs and Institutions in Groundwater Management, Water Supply and Health Sectors; Review. (Team Leader, David Sutherland; Consultant, Shireen Akhter) Report prepared under SSTA No. 4170-BAN: Arsenic Mitigation Review and Strategy Formulation.

Aziz, K. M. Ashraful, and Clarence Maloney
1985 Life Stages, Gender and Fertility in Bangladesh. Dhaka: International Centre for Diarrhoeal Disease Research, Bangladesh.

Baartmans, Frans
1990 Apah, the Sacred Waters: An Analysis of a Primordial Symbol in Hindu Myths. Delhi: B.R. Publishing

Corporation [Division of D.K. Publishers Distributors (P) Ltd.]

Bakker, Karen
2012 Commons Versus Commodities; Debating the Human Right to Water. *In* The Right to Water; Politics, Governance and Social Struggles. Farhana Sultana and Alex Loftus, eds. pp. 19-44. London and New York: Earthscan.

Barenstein, Jennifer Duyne
2008 Endogenous Water Resource Management in North-East Bangladesh; Lessons from the Haor Basin. *In* Water First; Issues and Challenges for Nations and Communities in South Asia. Kuntala Lahiri-Dutt and Robert J. Wasson, eds. pp. 349-371. Los Angeles, London, New Delhi, Singapore: Sage Publications India Pvt Ltd.

Basham, A. L.
1959 The Wonder That Was India; A Survey of the Culture of the Indian Sub-continent Before the Coming of the Muslims. New York : Grove Press.

Bateman, O. Massee, Raquiba A. Jahan, Sumana Brahman, Sushila Zeitlyn, and Sandra L. Laston
1995 Prevention of Diarrhea Through Improving Hygiene Behaviors; The Sanitation and Family Education (SAFE) Pilot Project Experience. ICDDR,B Special Publication No. 42; EHP Joint Publication No. 4. Dhaka: Co-published by International Centre for Diarrhoeal Research, Bangladesh, CARE Bangladesh,

and Environmental Health Project, U.S. Agency for International Development. (Reprinted and posted online: http://www.ehproject.org/PSF/Joint_Publications/JP004SAFEr.pdf, accessed November 2013)

Bhattacharyya, Narendra Nath
1996 Ancient Indian Rituals and Their Social Contents. Original Delhi: Manohar.

Blanchet, Thérèse
1984 Women, Pollution and Marginality; Meanings and Rituals of Birth in Rural Bangladesh. Dhaka: University Press Limited.

Brisbane Declaration
2007 The Brisbane Declaration: Environmental Flows Are Essential for Freshwater Ecosystem Health and Human Well-being. Declaration of the 10th International River Symposium and International Environmental Flows Conference, Brisbane, 3-6 September.

Brugnach, Marcela, and Helen Ingram
2012 Rethinking the Role of Humans in Water Management: Toward a New Model of Decision-Making. *In* Water, Cultural Diversity, and Global Environmental Change; Emerging Trends, Sustainable Futures? Barbara Rose Johnston, Editor-in-Chief. Paris, Dordrecht, Heidelberg, London, New York: UNESCO-IHP and Springer. pp.49-64.

Bürgel, J. Christoph
1976 Secular and Religious Features of Medieval Arabic Medicine. *In* Asian Medical Systems: A Comparative Study, Charles Leslie, ed. Berkeley, Los Angeles, and London: University of California Press. pp. 44-62.

Bywater, Krista
2012 Anti-privatization Struggles and the Right to Water in India. *In* The Right to Water; Politics, Governance and Social Struggles. Farhana Sultana and Alex Loftus, eds. pp. 206-222. London and New York: Earthscan.

Caldwell, Bruce
2008 When a Public Health Story Goes Sour; Arsenic Contaminated Drinking Water in Bangladesh. *In* Water First; Issues and Challenges for Nations and Communities in South Asia. Kuntala Lahiri-Dutt and Robert J. Wasson, eds. pp. 142-159. Los Angeles, London, New Delhi, Singapore: Sage Publications India Pvt Ltd.

CARE Bangladesh
2001 Sanitation and Family Education Resource (SAFER) Project; Report on Final Evaluation.

Chambers, Robert
1991 Shortcut and Participatory Methods for Gaining Social Information for Projects. *In* Putting People First; Sociological Variables in Rural Development. Michael M. Cernea, ed. pp. 515-537. New York: published for the World Bank by Oxford University Press (second edition).

Chowdhury, Uttam K., et al.

2000 Groundwater Arsenic Contamination in Bangladesh and West Bengal, India. Environmental Health Perspectives 108(5):393-397.

Crooke, W.

1894 An Introduction to the Popular Religion and Folklore of Northern India. Allahabad: Government Press, North-Western Provinces and Oudh.

1896 The Popular Religion and Folk-lore of Northern India (2 volumes). Delhi: Munshiram Manoharlal. Second edition.

Crow, Ben, and Farhana Sultana

2002 Gender, Class, and Access to Water: Three Cases in a Poor and Crowded Delta. Society and Natural Resources 15:709-724.

Dimock, Edward C., editor and translator

1963 The Thief of Love; Bengali Tales from Court and Village. Chicago: The University of Chicago Press.

Disanayaka, J.B.

1984 Aspects of Sinhala Folklore. Colombo: Lake House Investments Ltd.

2000 Water Heritage of Sri Lanka. Colombo: Ministry of Mahawali Development, Government of Sri Lanka, second revised edition.

Douglas, Mary

1966 Purity and Danger; An Analysis of Concepts of

Pollution and Taboo. London: Routledge & Kegan Paul.

Dumont, Louis [Mark Sainsbury, trans.]
1970 Homo Hierarchicus; The Caste System and Its Implications. London: Paladin (1972 reprint).

Duyne, Jennifer E.
1998 Local Initiatives: People's Water Management Practices in Rural Bangladesh. Development Policy Review 16(3):265-280.
2004 Local Initiatives; Collective Water Management in Rural Bangladesh. New Delhi: D.K. Printworld (P) Ltd.

Eck, Diana
1987 Rivers. *In* The Encyclopedia of Religion, Mircea Eliade et al., eds., Vol. 12. pp. 425-428 New York: Macmillan.
2012 India; A Sacred Geography New York: Harmony Publishers.

Eidsvik, Erlend
2003 Semiotics of a Sacred River. Signs and Meaning Ascribed [to] the Holy River Bagmati. Paper presented at International Water History Association 3rd International conference, Alexandria, Egypt, December 2003.

Ellickson, Jean
1972a A Believer Among Believers: The Religious Beliefs, Practices, and Meanings in a Village in Bangladesh.

Michigan State University, Ph.D. thesis. Ann Arbor, Michgan: University Microfilms.

1972b Symbols in Muslim Bengali Family Rituals. *In* Prelude to Crisis: Bengal and Bengal Studies in 1970. Peter J. Bertocci, ed. pp. 65-77. East Lansing: Asian Studies Center, Michigan State University.

Elmendorf, Mary

1981 Women, Water and the Decade. WASH Technical Report No. 6, prepared for USAID, DS/HEA; ODT No. 35. Presented at the American Water Works Association, 1981 Centennial Conference, International Affairs Session, St. Louis, MO.

Ennis-McMillan, Michael C.

2006 A Precious Liquid; Drinking Water and Culture in the Valley of Mexico. Belmont, CA: Thomson Wadsworth

Fagan, Brian

2011 Elixir; A History of Water and Humankind. New York: Bloomsbury Press.

Fairservis, Walter A., Jr.

1979a The Harappan Civilization: New Evidence and More Theory. *In* Ancient Cities of the Indus. Gregory L. Possehl, ed. pp. 49-65. Durham, North Carolina: Carolina Academic Press.

1979b The Origin, Character and Decline of an Early Civilization. *In* Ancient Cities of the Indus. Gregory L. Possehl, ed. pp. 66-89. Durham, North Carolina: Carolina Academic Press.

Faruqui, Naser I., Asit K. Biswas, and Murad J. Bino, eds.
2001 Water Management in Islam. Tokyo; New York: United Nations University Press.

Feldhaus, Anne
1995 Water and Womanhood; Religious Meanings of Rivers in Maharashtra. New York & Oxford: Oxford University Press.
2003a Connected Places: Region, Pilgrimage, and Geographical Imagination in India. New York: Palgrave Macmillan.
2003b Water Lore. *In* South Asian Folklore; An Encyclopedia. Margaret A. Mills, Peter J. Claus, and Sarah Diamond, eds. p. 634 New York & London: Routledge.

Foster, George M.
1976 Disease Etiologies in Non-western Medical Systems. American Anthropologist 78(4):773-782.
1987 On the Origin of Humoral Medicine in Latin America. Medical Anthropology Quarterly 1(4):355-393.
1994 Hippocrates' Latin American Legacy; Humoral Medicine in the New World. Langhorne, PA: Gordon and Breach.

Frazer, James George (Theodor H. Gaster, editor)
1959 The New Golden Bough. Garden City, NY: Anchor Books, Doubleday & Company.

Gender and Water Alliance
2002 Gender Water Management; Lessons Learnt Around the Globe; Findings of an electronic conference series convened by the Gender and Water Alliance.

Global Water Partnership
2000 Integrated Water Resources Management. Stockholm: Global Water Partnership. Technical Advisory Committee Paper No. 4.

Goffman, Erving
1968 Stigma: Notes on the Management of Spoiled Identity. Hammondsworth: Penguin Books.

Gomes, Bernadette Maria
2005 Vosaad: The Socio-Cultural Force of Water (A Study from Goa). Sociological Bulletin 54(2):250-276.

Government of Bangladesh, Ministry of Foreign Affairs, and DANIDA
1998 Five Districts Water Supply and Sanitation Group; Baseline Survey (Suzanne Hanchett, Team Leader). Vol. 1, Summary and Recommendations. Dhaka: DHV Consultant BV, Aqua Consultants & Associates Ltd., DEVCONsultants Ltd. (Danida reference no.: J.nr.104 Bang.174)
1999 Five Districts Water Supply and Sanitation Group; Baseline Survey (Suzanne Hanchett, Team Leader). Part 2. Dhaka. DHV Consultant BV, Aqua Consultants & Associates Ltd., DEVCONsultants Ltd. (Danida reference no.: J.nr.104 Bang.174)

Groenfeldt, David
2013 Water Ethics; A Values Approach to Solving the Water Crisis. London and New York: Routledge.

Hanchett, Suzanne

1975 Hindu Potlatches: Ceremonial Reciprocity and Prestige in Karnataka. *In* Competition and Modernization in South Asia. Helen E. Ullrich, ed. pp. 27-59. New Delhi: Abhinav.

1988 Coloured Rice; Symbolic Structure in Hindu Family Festivals. Delhi: Hindustan Publishing Corp., Ltd.

2003 Life Cycle Rituals. In South Asian Folklore; An Encyclopedia. Margaret A. Mills, Peter J. Claus, and Sarah Diamond, eds. pp. 354-358. New York & London: Routledge.

2006 Social Aspects of the Contamination of Drinking Water: A Review of Knowledge and Practice in Bangladesh and West Bengal. *In* Selected Papers on the Social Aspects of Arsenic and Arsenic Mitigation in Bangladesh. pp. 1-51. Dhaka, Bangladesh: Arsenic Policy Support Unit, Ministry of Local Government Rural Development and Cooperatives, Government of Bangladesh. Available online at: http://phys4.harvard.edu/~wilson/arsenic/references /selected_social_papers.pdf.

2009 Statement for International Learning Exchange: Arsenic in Bangladesh, organized by UNICEF Bangladesh and the Department of Public Health Engineering, 6 April 2009. Ms., 4 pages (available online: http://www.planningalternatives.com/id31.html).

2011 Programs and Pollution: Establishing Universal Sanitation Coverage in Rural Bangladesh. Presentation to Society for Applied Anthropology annual meetings. (Available online at http://www.planningalternatives.com/id18.html)

2012a A New Type of 'Social Problem'. *In* Water, Cultural Diversity, and Global Environmental Change; Emerging Trends, Sustainable Futures? Barbara Rose Johnston, Editor-in-Chief. p. 249 Paris, Dordrecht, Heidelberg, London, New York: UNESCO-IHP and Springer. p. 249.

2012b Bengali Women's Ideas About Water Quality. Presentation to UN Conference on Sustainable Development, Rio de Janeiro, Brazil. (Available at http://www.planningalternatives/id18.html)

forthcoming 2015 Sustainability and Society. *In* Envisioning Sustainabilities in Times of Disaster, Fiona Murphy and Pierre McDonagh, eds. Newcastle Upon Tyne, UK: Cambridge Scholars Publishing.

Hanchett, Suzanne, Kazi Rozana Akhter, Shireen Akhter, and Anwar Islam

2009 Meanings of Water in Bengali Culture. Paper presented at Fifth World Water Forum, Istanbul, Turkey, March 2009. (Available online at http://www.planningalternatives.com/id18.html)

Hanchett, Suzanne, Qumrun Nahar, Astrid van Agthoven, Cindy Geers, and Md. Ferdous Jamil Rezvi

2000 Arsenic Awareness in Six Bangladesh Towns. Dhaka: Royal Netherlands Embassy.

2002 Increasing Awareness of Arsenic in Bangladesh: Lessons from a Public Education Programme. Health Policy and Planning 17(4)393-401.

Hanchett, Suzanne, and Tofazzel Hossain Monju

2009 The Bangladesh Arsenic Mitigation and Water Supply

Project: A Public Administration and Public Health Failure. Paper presented at the annual meetings of the American Anthropological Association, Philadelphia, Pennsylvania, December 2009.

2012 Water in Rural Bangladesh: Governance and Social Issues. Presentation, annual meetings of the American Anthropological Association, San Francisco.

Hanchett, Suzanne, Jesmin Akhter, and Kazi Rozana Akhter

1998 Gender and Society in Bangladesh's Flood Action Plan. *In* Water, Culture, and Power. John M. Donahue and Barbara Rose Johnston, eds. pp. 209-234. Washington DC and Covelo CA: Island Press.

Handoo, J., and R. Kvideland, eds.

1999 Folklore: New Perspectives. Mysore: Zooni Publications.

Hartmann, Betsy, and James K. Boyce

1983 A Quiet Violence; View from a Bangladesh Village. Dhaka: University Press Ltd.

Hassan, M. Manzurul, Peter J. Atkins, and Christine E. Dunn

2005 Social Implications of Arsenic Poisoning in Bangladesh. Social Science and Medicine 61(10):2201-2211.

Hastrup, Kirsten

2009 Waterworlds: Framing the Question of Resilience. *In* The Question of Resilience; Social Responses to Climate Change. Kirsten Hastrup, ed. pp. 11-30. Copenhagen: The Royal Danish Academy of Sciences and Letters.

2013 Water and the Configuration of Social Worlds:

An Anthropological Perspective. Journal of Water Resources and Protection 5:59-66.

Hiwasaki, Lisa
2012 'Water for Life'...Water for Whose Life? Water, Cultural Diversity and Sustainable Development in the United Nations. *In* Water, Cultural Diversity, and Global Environmental Change; Emerging Trends, Sustainable Futures? Barbara Rose Johnston, Editor-in-Chief. pp. 509-531. Paris, Dordrecht, Heidelberg, London, New York: UNESCO-IHP and Springer.

Howes, Michael, and Robert Chambers
1979 Indigenous Technical Knowledge: Analysis, Implications and Issues. IDS Bulletin 10.2 Sussex: Institute for Development Studies

Inden, Ronald B., and Ralph W. Nicholas
1972 A Cultural Analysis of Bengali Kinship. *In* Prelude to Crisis: Bengal and Bengal Studies in 1970. Peter J. Bertocci, ed. pp. 91-97. East Lansing: Asian Studies Center, Michigan State University.
1977 Kinship in Bengali Culture. Chicago and London: University of Chicago Press.

Islam, Anwar
2012 Development, Disaster and Myth: A Case Study. *In* Water, Cultural Diversity, and Global Environmental Change; Emerging Trends, Sustainable Futures? Barbara Rose Johnston, Editor-in-Chief. p. 249 Paris, Dordrecht, Heidelberg, London, New York: UNESCO-IHP and Springer. p. 200.

Islam, Anwar, and Shireen Akhter

2012 Water Culture Traditions: Pond Myths. *In* Water, Cultural Diversity, and Global Environmental Change; Emerging Trends, Sustainable Futures? Barbara Rose Johnston, Editor-in-Chief. p. 249 Paris, Dordrecht, Heidelberg, London, New York: *UNESCO*-IHP and Springer. p. 189.

Islam, Mahmuda

1985 Women, Health and Culture. Dhaka: Women for Women.

Iyer, Ramaswamy R.

2001 Water: Charting a Course for the Future - II. Economic and Political Weekly, April 14, 2001. pp. 1235-1245.

2003 Water: Perspectives, Issues, Concerns. New Delhi and Thousand Oaks: Sage Publications.

Jakariya, Md.

2003 The Use of Alternative Safe Water Options to Mitigate the Arsenic Problem in Bangladesh: Community Perspective. Research Monograph Series No. 24. Dhaka: BRAC, Research and Evaluation Division.

James, Diana

2006 Re-Sourcing the Sacredness of Water. *In* Fluid Bonds; Views on Gender and Water. Kuntala Lahiri-Dutt, ed. pp. 85-104. Canberra: The National Institute for Environment (NIE), The Australian National University.

Jashim Uddin, Mohammed, and Newaz Ahmed Chowdhury
1999 Common Property Resources and Its Management; A Study on Water bodies. Kotbari, Comilla: Bangladesh Academy for Rural Development (BARD)

Jellife, D. B.
1957 Social Culture and Nutrition: Cultural Blocks and Protein Malnutrition inEarly Childhood in Rural West Bengal. Pediatrics 20:128-138.

Johnston, Barbara Rose, Editor-in-Chief
2012 Water, Cultural Diversity and Global Environmental Change: Emerging Trends, Sustainable Futures? Paris, Dordrecht, Heidelberg, London, New York: UNESCO-IHP and Springer.

Johnston, Richard Bart, Suzanne Hanchett, and Mohidul Hoque Khan
2010 The Socio-economics of Arsenic Removal. Nature Geoscience 3(1):2-3.

Kabir, Humayun
1947 Men and Rivers. London: New India Publishing Company.

Kleinman, Arthur
1978 Concepts and a Model for the Comparison of Medical Systems as Cultural Systems. Social Science and Medicine 12:85-93.

Kränzlin, Irène
2000 Pond Management in Rural Bangladesh: System

Changes, Problems and Prospects, and Implications for Sustainable Development. Basel: Wepf.

Krishnamurthy, Radha
1996 Water in Ancient India. Indian Journal of History of Science 31(4):327-337.

Kumar, Krishna, ed.
1993 Rapid Appraisal Methods. Washington, DC: The World Bank.

Kumar, Savitri V.
1983 The Pauranic Lore of Holy Water-Places: with Special Reference to Skanda Purana. New Delhi: Munshiram Manoharlal.

Lahiri-Dutt, Kuntala
2006 Nadi O Nari: Representing the River and Women of the Rural Communities in the Bengal Delta. *In* Fluid Bonds; Views on Gender and Water, Kuntala Lahiri-Dutt, ed. pp. 387-408. Canberra: The National Institute for Environment (NIE) & Australian National University.

Lahiri-Dutt, Kuntala, ed.
2006 Fluid Bonds; Views on Gender and Water. Canberra: The National Institute for Environment (NIE), Australian National University.

Lahiri-Dutt, Kuntala, and Robert J. Wasson, eds.
2008 Water First; Issues and Challenges for Nations and

Communities in South Asia. Los Angeles, London, New Delhi, Singapore: Sage Publications.

Lahiri-Dutt, Kuntala, and Gopa Samanta
2013 Dancing with the River; People and Life on the Chars of South Asia. New Haven, CT, and London: Yale University Press.

Lapidus, Ira M.
2012 Islamic Societies to the Nineteenth Century; A Global History. Cambridge and New York: Cambridge University Press.

Leone, Faye
2014 The Pre-2015 Agenda: Status of Efforts to Devise the Post-2015 Development Agenda. International Institute for Sustainable Development Reporting Services, Policy Updates #1. (Available online at http://post2015.iisd.org/policy-updtes/the-pre-2015-agenda-status-of-efforts-to-devise-the-post-2015-development-agenda/)

Leslie, Charles, ed.
1976 Asian Medical Systems: A Comparative Study. Berkeley, CA: University of California Press.

Linton, Jamie
2006 The Social Nature of Natural Resources - The Case of Water. Reconstruction: Studies in Contemporary Culture 6.3 Document available online: http://reconstruction.eserver.org/063/linton.shtml
2010 What Is Water? History of a Modern Abstraction. Vancouver and Toronto: University of British Columbia Press.

Mahapatra, Piyush Kanti

1963 Some Rain Ceremonies of West Bengal. *In* Rain in Indian Life and Lore. Sankar Sen Gupta, ed. pp. 63-66. Calcutta: Indian Publications.

Maity, Pradyot Kumar

1966 Historical Studies in the Cult of the Goddess Manasa; A Socio-Cultural Study. Calcutta: Punthi Pustak.

1971 Popular Cults, Legends and Stories in Ancient Bengal. Calcutta: Punthi Pustak and Indian Publications.

1988 Folk-Rituals of Eastern India. New Delhi: Abhinav Publications.

Maloney, Clarence, K.M. Ashraful Aziz, and Profulla C. Sarker

1981 Beliefs and Fertility in Bangladesh. Rajshahi: Institute of Bangladesh Studies, University of Rajshahi. (Also Dacca: International Centre for Diarrhoeal Disease Research, Bangladesh. Monograph No. 2)

Milton, Abul Hasnat, Ziaul Hasan, S.M. Shahidullah, Sinthia Sharmin, M.D. Jakariya, Mahfuzar Rahman, Keith Dear, and Wayne Smith

2004 Association Between Nutritional Status and Arsenicosis Due to Chronic Arsenic Exposure in Bangladesh. International Journal of Environmental Health Research 14(2):99-108.

Morgan, Monica

2012 Cultural Flows: Asserting Indigenous Rights and Interests in the Waters of the Murray-Darling River System, Australia. *In* Water, Cultural Diversity, and Global Environmental Change; Emerging Trends,

Sustainable Futures? Barbara Rose Johnston, Editor-in-Chief. Paris, Dordrecht, Heidelberg, London, New York: UNESCO-IHP and Springer. pp. 453-466.

Mosse, David

2003 The Rule of Water; Statecraft, Ecology and Collective Action in South India. New Delhi: Oxford University Press.

Nasreen, Mahbuba

2003 Social Impact of Arsenicosis. *In* Arsenic Contamination: Bangladesh Perspective. M. Feroze Ahmed, ed. pp. 340-353. Dhaka: ITN-Bangladesh University of Engineering and Technology (BUET).

2008 Coping with Arsenicosis: Socio-economic and Gender Response to Arsenic Contamination of Ground Water in Bangladesh. Research report prepared with financial support from ITN-BUET. 15 page ms. Dhaka: Bangladesh University of Engineering and Technology.

Nichter, Mark

1985 Drink Boiled Water: A Cultural Analysis of a Health Education Message. Social Science and Medicine 21(6):667-669.

1987 Cultural Dimensions of Hot, Cold and Sema in Sinhalese Health Culture. Social Science and Medicine 25:377-387.

Nichter, Mark, ed.

1992 Anthropological Approaches to the Study of Ethnomedicine. Langhorne, PA: Gordon and Breach Science Publishers.

Obeysekere, Gananath

1976 The Impact of Āyurvedic Ideas on the Culture and the Individual in Sri Lanka. *In* Asian Medical Systems: A Comparative Study, Charles Leslie, ed. pp. 201-226. Berkeley and Los Angeles: University of California Press.

1978 Illness, Culture, and Meaning: Some Comments on the Nature of Traditional Medicine. *In* Culture and Healing in Asian Societies; Anthropological, Psychiatric and Public Health Studies, Arthur Kleinman, Peter Kunstadter, E. Russell Alexander, and James L. Gate, eds. pp. 253-264. Cambridge, MA: Schenkman Publishing Co.

Orlove, Ben, and Steven C. Caton

2009 Water as an Object of Anthropological Inquiry. *In* The Question of Resilience; Social Responses to Climate Change. Kirsten Hastrup, ed. pp. 31-47. Copenhagen: The Royal Danish Academy of Sciences and Letters.

2010 Water Sustainability: Anthropological Perspectives and Prospects. Annual Review of Anthropology 39:401-415. Palo Alto, CA: Annual Reviews.

Ortner, Sherry

2011 Purification Rite. Encyclopaedia Brittanica Online. (Retrieved from http://www.brittanica.com/EBchecked/topic/483975/purification-rite)

Papayannis, Thymio, and Dave Pritchard

2008 Culture and Wetlands: A Ramsar Guidance Document. Gland, Switzerland: Ramsar Convention.

2010 Wetland Cultural and Spiritual Values, and the Ramsar

Convention. *In* Sacred Natural Sites; Conserving Nature and Culture. Bas Verschuuren, Robert Wild, Jeffrey McNeely, and Gonzalo Oviedo, eds. pp. 180-187. London & Washington DC: Earthscan.

Planning Alternatives for Change, LLC, and Pathways Consulting Services Ltd.

2003 A UNICEF-WHO Joint Project on Building Community Based Arsenic Mitigation Response Capacity in Bhanga, Muradnagar, and Serajdikhan, Bangladesh. Report (midterm report) submitted to World Health Organization Bangladesh. Dhaka.

2006 Final Evaluation Report on Building Community Based Arsenic Mitigation Response Capacity in Muradnagar, Serajdikhan and Bhanga Upazilas of Bangladesh. Report submitted to World Health Organization Bangladesh, Water and Environmental Section (under Contract No. BAN PHE 003XW 04). Dhaka.

2009 Final Summary Report on Social and Economic Assessment of Arsenic Removal Technologies. Submitted to Water and Environmental Sanitation Section, UNICEF Bangladesh.

Poff, N.L., et al.

2010 The Ecological Limits of Hydrologic Alteration (ELOHA): A New Framework for Developing Regional Environmental Flow Standards. Freshwater Biology 55:147-170.

Rahman, Mahmud

2010 Killing the Water: Stories. New York: Penguin Books.

Rashid, Haroun Er

1978 Geography of Bangladesh. Boulder, Colorado: Westview Press.

1991 Geography of Bangladesh. Dhaka: University Press Limited, second revised edition.

Rau, Shanta Rama

1962 Introduction. *In* Chemmeen, a novel by Thakazhi Sivasankara Pillai (Narayana Menon, translator). Bombay, New Delhi, Calcutta, Madras: Jaico Publishing House.

Ravenscroft, Peter, Hugh Brammer, and Keith Richards

2009 Arsenic Pollution: A Global Synthesis. West Sussex: Wiley-Blackwell.

Risley, H.H.

1892 Tribes and Castes of Bengal; Ethnographic Glossary; Vol. 1. Calcutta: Printed at the Bengal Secretariat Press.

Rizvi, Najma

1979 Rural and Urban Food Behavior in Bangladesh: An Anthropological Perspective to the Problem of Malnutrition. Los Angeles, CA: University of California Los Angeles, Ph.D. dissertation

1986 Food Categories in Bangladesh and Its Relationship to Food Beliefs and Practices of Vulnerable Groups. *In* Food, Society, and Culture: Aspects in South Asian Food Systems. R.S. Khare and M.S.A. Rao, eds. pp. 223-251. Durham, NC: Carolina Academic Press.

Rocke, Alan J.
2012 History of Science. Encyclopaedia Brittanica, online edition.

Rosenboom, Jan Willem
2004 Not Just Red or Green: An Analysis of Arsenic Data from 15 Upazilas in Bangladesh. Dhaka: UNICEF Bangladesh, Water and Environmental Sanitation Unit.

Rosin, R. Thomas
1993 The Tradition of Groundwater Irrigation in Northwestern India. Human Ecology 21(1):51-86.

Rosman, Abraham, and Paula Rubel
1981 The Tapestry of Culture; An Introduction to Cultural Anthropology. Glenview, Illinois: Scott Foresman and Company (first edition).

Ross, Anne, Kathleen Pickering Sherman, Jeffrey G. Snodgrass, Henry D. Delcore, and Richard Sherman
2011 Indigenous Peoples and the Collaborative Stewardship of Nature; Knowledge Binds and Institutional Conflicts. Walnut Creek, California: Left Coast Press, Inc.

Schelwald-van der Kley, Lida, and Linda Reijerkerk
2009 Water: A Way of Life; Sustainable Water Management in a Cultural Context. Boca Raton, FL & London: CRC Press, Taylor & Francis Group (A Balkema Book).

Sen Gupta, Sankar, ed.
1963 Rain in Indian Life and Lore. Calcutta: Indian Publications.

Shamim, Ishrat, and Khaleda Salahuddin

1994 Energy and Water Crisis in Rural Households: Linkages with Women's Work and Time. Dhaka: Women for Women.

Shiva, Vandana

2002 Water Wars; Privatization, Pollution and Profit. Cambridge, MA: South End Press.

2004 Building Water Democracy: People's Victory Against Coca-Cola in Plachimada. New Delhi: Research Foundation for Science, Technology, and Ecology.

Siddiqui, Habib

n.d. The Soul of Hajj. (Online document downloaded from: http://www.islamicity.com/articles/Articles.asp?ref=IC0401-2197#sthash.fJZjJVz5.dpbs)

Smith, Allan, Elena O. Lingas, and Mahfuzar Rahman

2000 Contamination of Drinking-water by Arsenic in Bangladesh: A Public Health Emergency. Bulletin of the World Health Organization 78(9):1093-1103.

Strang, Veronica

2004 The Meaning of Water. Oxford and New York: Berg.

Sultana, Farhana

2006a Gender Concerns in Arsenic Mitigation in Bangladesh: Trends and Challenges. *In* Selected Papers on the Social Aspects of Arsenic and Arsenic Mitigation in Bangladesh. pp. 53-84. Dhaka, Bangladesh: Arsenic Policy Support Unit, Ministry of Local Government Rural Development and Cooperatives, Government of Bangladesh (available on-line: http://phys4.harvard.edu/~wilson/arsenic/references/selected_social_papers.pdf).

2006b Gendered Waters, Poisoned Wells: Political Ecology of the Arsenic Crisis in Bangladesh. *In* Fluid Bonds; Views on Gender and Water. Kuntala Lahiri-Dutt, ed. pp. 362-386. Canberra: The National Institute for Environment (NIE), The Australian National University.

2009 Fluid Lives: Subjectivities, Gender and Water in Rural Bangladesh. Gender, Place and Culture 16(4):427-44.

Sultana, Farhana, and Alex Loftus, eds.

2012 The Right to Water: Politics, Governance and Social Struggles. Earthscan, Water Text Series.

Tharme, R.E.

2003 A Global Perspective on Environmental Flow Assessment: Emerging Trends in the Development and Application of Environmental Flow Methodologies for Rivers. River Research and Applications 19(5-6):397-441.

Thornton, Thomas F.

2012 Watersheds and Marinescapes: Understanding and Maintaining Cultural Diversity Among Southeast Alaska Natives. *In* Water, Cultural Diversity, and Global Environmental Change; Emerging Trends, Sustainable Futures? Barbara Rose Johnston, Editor-in-Chief. pp. 123-136. Paris, Dordrecht, Heidelberg, London, New York: UNESCO-IHP and Springer.

Tipa, G., and K. Nelson

2008 Introducing Cultural Opportunities: A Framework for Incorporating Cultural Perspectives in Contemporary Resource Management. Journal of Environmental Policy and Planning 10:313-337.

Tomizawa, Hitomi
2001 Arsenic Poisoning from Ground Water in Rural Bangladesh: The Potential for Women's Participatory Environmental Education. Wageningen: Wageningen University, Social Sciences, M.Sc. thesis.

Tratschin, Risch
n.d. Water, Sanitation and Culture. Online document: http://sswm.info/content/water-sanitation-and-culture, accessed 30 November 2013.

Uddin, Mohammed Jashim, and Newaz Ahmed Chowdhury
1999 Common Property Resources and Its Management; A Study on Water bodies. Kotbari, Comilla: Bangladesh Academy for Rural Development.

UN-Water
2014 A Post-2015 Global Goal for Water: Synthesis of key findings and recommendations from UN-Water. Executive Summary; Detailed Targets and Associated Indicators.

UNESCO
1982 Mexico City Declaration on Cultural Policies. Mexico City: UNESCO.

UNICEF and World Health Organization
2009 Progotir Pathey. Dhaka.

UNIDO
[2001] Concerted Action on Elimination/Reduction of Arsenic in Ground Water, West Bengal, India (A.K. Sengupta, author).

United Nations, Department of Economic and Social Affairs
2004 The Concept of Indigenous People. Background paper prepared by the Secretariat of the Permanent Forum on Indigenous Issues.

United Nations General Assembly, Human Rights Council
2013 Report of the Special Rapporteur on the Human Right to Safe Drinking Water and Sanitation, Catarina Albuquerque. [New York:] Twenty-fourth session, Document No. A/HRC/24/44.

United Nations Permanent Forum on Indigenous Issues
n.d. Indigenous Peoples, Indigenous Voices. Online document downloaded October 2013 from: http://www.un.org/esa/socdev/unpfii/documents/5session_factsheet1.pdf

Van Geen, Alexander, Kazi Matin Ahmed, and Joseph H. Graziano
2005 Cleaning Up Bangladesh's Deadly Wells. New York Times (Op. Ed.), Monday, August 1, 2005.

van Gennep, Arnold (Monika B. Vizedom and Gabrielle L. Caffee, translators)
1960 The Rites of Passage. Chicago, IL: The University of Chicago Press, Phoenix Books.

van Wijk, C.
1998 Gender in Water Resources Management, Water Supply and Sanitation: Roles and Realities Revisited. The Hague: IRC, Technical Paper No. 33-E.

World Bank
2007 Implementation Completion and Results Report (IDA-31240 SWTZ-21082) on a Credit in the Amount

of Sdr 24.2 Million (Us$ 44.4 Million Equivalent) to Bangladesh for Arsenic Mitigation Water Supply (Report No: ICR000028). Washington, DC: The World Bank. (Available online at http://www.planningalternatives.com/id31/html)

Zeitlyn, Sushila, and Farzana Islam
1990 Patterns of Child Feeding and Health Seeking Behaviour in Bangladesh: Two Case Studies. Dhaka: ICDDR,B Monograph.

Zeitlyn, Sushila, G.N. Faisal, and K.M. Ashraful Aziz, and T. Sharmin
1993 Women and Health: Exploring the Socio-cultural Barriers and Determinants of Women's Health Status in Rural Bangladesh [ICDDR,B Protocol]. Dhaka: International Centre for Diarrhoeal Diseases Research, Bangladesh & Jahangirnagar University

Zimmer, Heinrich
1972 Myths and Symbols in Indian Art and Civilization. Princeton, NJ: Princeton University Press (Bollingen Series VI).

Appendix 1. Principal Study Locations

District (Year)	Subdistricts & Towns/Unions	Topics of Study	Project Assignments
Bangladesh			
Barguna (1997-98)	Pathargatha Town Bamna Town	Water sources used Perceptions of water quality	Baseline study (S. Hanchett, Shireen Akhter, Anwar Islam)
Comilla (2003, 2006,2009)	Homna: Ghar-mora and Chander Char unions	General study	UNICEF: Water Culture Study (All authors)
	Muradnagar: Muradnagar and Poschim Nabipur unions	Arsenic problem and alternative solutions	UNICEF: Evaluation studies of arsenic mitigation program (All authors)
	Laksham: Gobindapur, Madaffar-ganj, Purbo Laksham, and other Laksham unions	Arsenic problem and alternative solutions	

District (Year)	Subdistricts & Towns/Unions	Topics of Study	Project Assignments
Laksmipur (1997-98, 2006)	Raipur Town	Water sources used Perceptions of water quality	Baseline study
Laksmipur (2006)	Ramganj and others	Arsenic problem and mitigation efforts	Evaluation study of BAMWSP for World Bank (All authors)
Noakhali (1997-98 & 2005-10)	Noakhali City/ Pourashava (squatter settlements, other locations) Chatkhil Town	Water sources used Perceptions of water quality	Baseline study (S. Hanchett, S. Akhter, A. Islam)
	Hatiya	Women's participation in water and sanitation activities Development program for poor people	Char Development Sector Project (S. Akhter)

District (Year)	Subdistricts & Towns/Unions	Topics of Study	Project Assignments
Pabna	Bera Subdistrict, Ruppur Union	General study UNICEF: Water Culture Study & Arsenic problem and solutions	UNICEF: Evaluation study of social acceptability of arsenic removal filters (All authors)
Patuakhali (1997-98)	Patuakhali Pourashava [Urban area] Bauphal, Dasmina, Kalapara, Mirzaganj, Galachipa towns	Water sources used Perceptions of water quality	Baseline study
Satkhira (2009-10)	Shyamnagar Village, Garbura Union	Water governance Salinity Post-cyclone recovery Climate change	Progothi, an NGO working on drinking water (T. Hossain Monju)
Tangail (since 1980s)	Delduar	Multiple topics	Anwar Islam's residence

District (Year)	Subdistricts & Towns/Unions	Topics of Study	Project Assignments
West Bengal, India			
North 24 Parganas & Kolkata City (2004)	Deganga	Arsenic testing and mitigation Key informant interviews	Arsenic Policy Support Unit study of social aspects of the arsenic problem (S. Hanchett, S. Akhter)

Appendix 2. Water Vocabulary

Bengali Word or Expression: Translitera-tion (Bengali Script)	Translation	Informa-tion Source (*District/ Subdistrict)	Other
*1-Laksham, 2-Homna, 3-Bera, 4-Noakhali, Feni, &/or Laksmipur, 5-Delduar, 6-Patuakhali &/or Barguna			
1. General Words			
paani (পানি)	Water	1,2,3,4,5,6	Muslims' normal word
jal (জল)	Water	1,2,3,4,5,6	Hindus' normal word
haani (হানি)	Water	1, 4	
niraapader paani nirapod paani (নিরাপদ পানি নিরাপদ)	Safe water		Common term used in all parts of Bangladesh

Bengali Word or Expression: Translitera-tion (Bengali Script)	Translation	Informa-tion Source (*District/ Subdistrict)	Other
*1-Laksham, 2-Homna, 3-Bera, 4-Noakhali, Feni, &/or Laksmipur, 5-Delduar, 6-Patuakhali &/or Barguna			
osh (ওষ)	Morning dew	1,2,5	Common term
paataal paataaler paani (পাতাল, পাতালরে পানি)	Under-ground water	1,2	Common term
aarshonik mukto paani র্সনেকি মুক্ত পানি(র্সনেকি মুক্ত পানি) also: *aar-chonik mukto paani*	Arsenic-free water	1,2	Common term
aarshonik/ aarchonik jukto paani (আর্সনেকি যুক্ত পানি)	Arsenic-contaminat-ed water	1,2	Common term
2. Rivers			
nadii (নদী)	River (fem.)	1,2,3,4,5,6	Common term

Bengali Word or Expression: Transliteration (Bengali Script)	Translation	Information Source (*District/ Subdistrict)	Other
*1-Laksham, 2-Homna, 3-Bera, 4-Noakhali, Feni, &/or Laksmipur, 5-Delduar, 6-Patuakhali &/or Barguna			
nad (নদ)	River (masc.)	1,3	Used mostly in schools
gaaŋ (গাঙ)	River	1,2,3,4,5,6	
gaaŋer paani (গাঙের পানি)	River water	1,2,3,4,5,6	
gher paani (ঘের পানি)	River water	2,3	Lit. "water accessed from steps or platform at water's edge"
kumor paani (কুমোর পানি)	Deep place in river, perennial	5	
bil or *biler paani* (বিল - বিলের পানি)	Swamp or marshland, Oxbow lake, Water of oxbow lake	1,2,3,5	Common term

Bengali Word or Expression: Transliteration (Bengali Script)	Translation	Information Source (*District/ Subdistrict)	Other
*1-Laksham, 2-Homna, 3-Bera, 4-Noakhali, Feni, &/or Laksmipur, 5-Delduar, 6-Patuakhali &/or Barguna			
3. Water in Agricultural Fields & Other Rural Spaces			
haaorer paani (হাওররে পানি)	Water on low, marshy land or geological depression (*haaor*)	3	Common term
jolaar paani (জলার পানি)	Water in low-lying or marshy land	1,2,3,4,5	
khaapaaler paani (খাপালরে পানি)	Water in a marshy area	3	
gaaRaa (গাড়া)	Ditch with water in it, has round shape	2	
maiThaal (মাইঠাল)	Small, seasonal water body in a field	3	

Bengali Word or Expression: Translitera-tion (Bengali Script)	Translation	Informa-tion Source (*District/ Subdistrict)	Other
*1-Laksham, 2-Homna, 3-Bera, 4-Noakhali, Feni, &/or Laksmipur, 5-Delduar, 6-Patuakhali &/or Barguna			
goR (গড়) *goRaa* (গড়া)	Natural drain to fields in rainy sea-son	1,2,5	
khaad (খাদ)	Ditch with water in it, bigger than *goR*	1,2	
do (দও)	Large, naturally occurring water body having eerily blue water	5	Considered ghostly and frightening
4. Ponds, Tanks, and Lakes			
pukur (পুকুর)	Pond	1,2,3,4,5,6	Common term
	Pond with steps (*ghaaT*)	1,3	
*pushkun*i (পুষ্কুনি)	Pond	1,2,3,4	

Bengali Word or Expression: Translitera-tion (Bengali Script)	Translation	Informa-tion Source (*District/ Subdistrict)	Other
*1-Laksham, 2-Homna, 3-Bera, 4-Noakhali, Feni, &/or Laksmipur, 5-Delduar, 6-Patuakhali &/or Barguna			
pushkundi (পুষ্কুন্দী) *pushkunDi* (পুষ্কুন্ডি)	Pond	3	
puhuir (পুহুইর) or *puhuirer paani* (পুহুইররে পানি)	Pond	1,2	
maiThaal (মাইঠাল) *maiTaal* (মাইটাল)	Very small, small or medium-sized pond, with no *ghaaT* (steps)	3,5	
dighi (দিঘি)	Very large, deep pond or lake, usually man-made	1,2,3,4,5,6	Common term usually referring to water-bodies established in the colonial and earlier periods

Bengali Word or Expression: Translitera-tion (Bengali Script)	Translation	Informa-tion Source (*District/ Subdistrict)	Other
*1-Laksham, 2-Homna, 3-Bera, 4-Noakhali, Feni, &/or Laksmipur, 5-Delduar, 6-Patuakhali &/or Barguna			
baaRir bhi-torer pu-kur (বাড়রি ভতিররে পুকুর)	"Inside" pond	1,2,4	Pond inside homestead (*baaRi*) boundary
baaRir baairer pu-kur (বাড়রি বাইররে পুকুর)	"Outside" pond	1,2,4	Pond outside homestead boundary
5. Water in Different States, Water in Motion, Water with Special Qualities			
chalti jal (চলতি জল)	Rapidly flowing river water	3	
Dholer paani (ঢলরে পানি)	Strongly, rapidly flowing water	1,2,3,5	Associated with heavy rains
notun paani (নতুন পানি)	Water of the first mon-soon	3	

Bengali Word or Expression: Translitera-tion (Bengali Script)	Translation	Informa-tion Source (*District/ Subdistrict)	Other
*1-Laksham, 2-Homna, 3-Bera, 4-Noakhali, Feni, &/or Laksmipur, 5-Delduar, 6-Patuakhali &/or Barguna			
noyaa paani (নয়া পানি)	New water; Water of the first mon-soon	1,2,4,5	
paani jo-bankaal (পানি যৌবনকাল)	Young water	1,2	
paani brid-dhokaal (পানি বৃদ্ধকাল)	Water in its old age	1,2	
baashi paani (বাসি পানি)	Old, stale water	3	
TolTolaa paani (টলটলা পানি)	Clear water, transparent-deep layer visible	3	
gholaa paani (ঘোলা পানি)	Turbid water or mixture of clean water and mud	1,2,3,4,5	

Bengali Word or Expression: Translitera-tion (Bengali Script)	Translation	Informa-tion Source (*District/ Subdistrict)	Other
*1-Laksham, 2-Homna, 3-Bera, 4-Noakhali, Feni, &/or Laksmipur, 5-Delduar, 6-Patuakhali &/or Barguna			
jharNaar paani (ঝর্নার পানি)	Under-ground water that comes up when a pond is dug	3	
	Water from a spring, fountain, waterfall	1,2	Holy to reli-gious Hindu people
jooaarer paani (জোয়ারের পানি)	Water flow-ing from upstream to downstream	1,2,3,4,5,6	
nimna chaaper paani (নিম্না চাপের পানি)	Tidal water flowing from the sea into the river	3,4,6	
gaaRaar paani (গাড়ার পানি)	Stagnant, dirty water in a low place	3	

Bengali Word or Expression: Translitera-tion (Bengali Script)	Translation	Informa-tion Source (*District/ Subdistrict)	Other
*1-Laksham, 2-Homna, 3-Bera, 4-Noakhali, Feni, &/or Laksmipur, 5-Delduar, 6-Patuakhali &/or Barguna			
nardo-maar paani (নর্দমার পানি)	Water in a hole, pit or drain	1,3	
garto (গর্ত)	Hole/pit, possibly containing waste water	1,3	Common term
6. Rain			
brishTi (বৃষ্টি)	Rain	1,2,3	Common term
brishTir paani (বৃষ্টির পানি)	Rain water	1,2,3	Common term
aasmaaner paani (আসমানের পানি)	Water from the sky	1,2,3	
megher paani (মেঘের পানি)	Water from clouds	1,3	
deeowaa (দেওয়া)	Rain	3	Lit. "God-given"

Bengali Word or Expression: Transliteration (Bengali Script)	Translation	Information Source (*District/ Subdistrict)	Other
*1-Laksham, 2-Homna, 3-Bera, 4-Noakhali, Feni, &/or Laksmipur, 5-Delduar, 6-Patuakhali &/or Barguna			
aallaar paani (আল্লার পানি)	Rain	1,3	Lit. "Allah's water"
barshaar paani (বর্ষার পানি)	Monsoon rains	1,3	
baishyaa maasher paani (বাইশ্যা মাসের পানি)	Water in the monsoon season (*barsha,* বর্ষা) months		
aashaaRi brishTi (আষাঢ়ি বৃষ্টি)	Rain in the month of Ashar (June-July, early monsoon period)		
aakaash chhiRaa (আকাশ ছিড়া) *chhidra* (ছিদ্র) *tyaaraa paani* (ত্যারা পানি)	Continuous rain lasting for some days	1,2,3	

Bengali Word or Expression: Translitera-tion (Bengali Script)	Translation	Informa-tion Source (*District/ Subdistrict)	Other
*1-Laksham, 2-Homna, 3-Bera, 4-Noakhali, Feni, &/or Laksmipur, 5-Delduar, 6-Patuakhali &/or Barguna			
abiraam brishTi (অবিরাম বৃষ্টি)	Continuous rain	5	Common term
aadal (আদল) *baadal* (বাদল)	Continuous heavy rain lasting 2- 3 days	3	
bhaaRii brishTi (ভাড়ী বৃষ্টি)	Heavy rain		Common term
baraak (বরাক)	Heavy rain falling in-cessantly	3	
gaab kaaTaabaraak (গাব কাটাবরাক)	Speedy rain	3	
haaput (হাপুত) *saaput* (সাপুত)	Rain with storm and strong winds	3	
guRiguRi brishTi (গুঁড়ি গুঁড়ি বৃষ্টি)	Light rain, drizzle		Common term

Bengali Word or Expression: Translitera-tion (Bengali Script)	Translation	Informa-tion Source (*District/ Subdistrict)	Other
*1-Laksham, 2-Homna, 3-Bera, 4-Noakhali, Feni, &/or Laksmipur, 5-Delduar, 6-Patuakhali &/or Barguna			
guRinnii brishTi (গুঁড়িন্নি বৃষ্টি)	Light rain, drizzle	3	
fis paaRaa brishTi (ফিসি পাড়া বৃষ্টি)	Light rain	3	
kaanaa megher paani (কানা মেঘের পানি)	Rain and sunshine occurring simulta-neously	3	
ilshe guRi brishTi (ইলশে গুঁড়ি বৃষ্টি)	Tiny drops of rain		Common term Lit. “hilsha fish scales rain”
7. Hail			
choyaa-choy-aa brishTi (চোয়া চোয়া বৃষ্টি)	Rain with hail	3	

Bengali Word or Expression: Translitera-tion (Bengali Script)	Translation	Informa-tion Source (*District/ Subdistrict)	Other
*1-Laksham, 2-Homna, 3-Bera, 4-Noakhali, Feni, &/or Laksmipur, 5-Delduar, 6-Patuakhali &/or Barguna			
shil (শিল), *shoraa* (শোরা), *shaairaa* (শাইরা), *paa-thorer paani* (পাথরের পানি)	Hail	1,2,3,5	
8. Floods			
banna (বন্যা)	Normal flood		Common term
bannayaar paani (বন্যার পানি)	Monsoon water in a river/canal	1,2,3,4,5	
bhoyaboho banna (ভয়াবহ বন্যা)	Extreme flood		Common term
9. Water from Technical Devices			
kaler paani (কলের পানি) *nalku-per paani* (নলকূপের পানি)	Tube well water	1,2,3,5	Common terms

Bengali Word or Expression: Translitera-tion (Bengali Script)	Translation	Informa-tion Source (*District/ Subdistrict)	Other
*1-Laksham, 2-Homna, 3-Bera, 4-Noakhali, Feni, &/or Laksmipur, 5-Delduar, 6-Patuakhali &/or Barguna			
chaap koler paani (চাপ কলরে পানি)	Tube well water	1,2,4	
taãraa paam-per paani (তাঁরা পাম্পরে পানি)	Tube well water from a "tara" pump	3	Common term
filTaarer paani	Water from a Sono arse-nic filter	3	
chaari paani (চারি পানি)	Water of a home-made filter	3	
haaujer paani (হাউজরে পানি) *Teper paani* (টপেরে পানি) *taagarer paani* (তাগাররে পানি)	Water from a neighbor-hood Sidko arsenic-re-moval plant	3	
kuiyer (কুইয়রে) or *kuaar paani* (কুয়ার পানি)	Dug well water	1,2,3	

Bengali Word or Expression: Translitera-tion (Bengali Script)	Translation	Informa-tion Source (*District/ Subdistrict)	Other
*1-Laksham, 2-Homna, 3-Bera, 4-Noakhali, Feni, &/or Laksmipur, 5-Delduar, 6-Patuakhali &/or Barguna			
indaaraar paani (ইন্দারার পানি)	Water from a ring well (large, covered dug well lined with bricks)	3,5	
10. Problem-solving and Health-promoting Water			
paani paRaa (পানি পড়া)			Common term
shaaraN paRaa paani (শারণ পড়া পানি)	Water that has absorbed Holy Quran words	1,2	Used to promote healing and in childbirth
aab jamjam paani (আব যমযম পানি) *jamjam paani* (যমযম পানি)	Water brought from Mecca by a person who has performed the Hajj pilgrimage	1,2	
mariyam fuler paani (মরিয়ম ফুলের পানি)	Mariam flower in water	1,2	

Appendix 3. Lyrics of a Bengali Rain-making Song

English Translation	Bengali
Oh Cloud, Our Cloud	*Aago, megh O*
Oh mother of Cloud	*Shua khuliir maago*
Drop rain continuously	*Eteer kulaa beeteer baaN*
On the winnowing-fan made of cane	*Jorjorii di jharii aan*
Oh Allah Cloud Cloud	*Allah megh O megh O*
Mother of Cloud died	*Megheer Maa morii gejhee*
	Haani haiboo koi
Where will we find her?	
The seven brother fishermen	*Jaollaara haat bai*
Going home to make their nets	*Jaal maartoo haani nai*
Water in the drain of *aurum* (taro)	*Kochu khetee naalaa haani*
Come Cloud Water	*Aairee megheer haani*
Drive away the brother. Send him that way	*Ei baairgor go eyaand di kaaTaa*
To that way	*Hiyaan di kaaTaa*
Sleepy, sleepy tiger cub	*Gumur gumur baa-gheer saa*
Give him rice and send him away	*Chail deen cholii jai*
	Haani deen bijii jai

Index

A

B

C

D

E

F

G

H

I

J

K

L

M

O

P

R

S

T

U

V

W

About the Authors

Suzanne Hanchett is a social anthropologist with a doctorate from Columbia University. She is a Partner in the consulting firm, Planning Alternatives for Change LLC and a Researcher at the Center for Political Ecology. She has done basic and applied research on social structure, gender, and poverty-related issues in Bangladesh, India, and several other countries. She is the author of *Coloured Rice*, a book on symbolism of home-centered Hindu myths and rituals in Karnataka State, India. Since 1997 she has been working with water and sanitation programs and arsenic mitigation projects. She has served as Team Leader for several applied studies conducted together with the other authors.

Tofazzel Hossain Monju was born in Pirojpur, a southern coastal district of Bangladesh. At age fourteen he joined the Bangladesh liberation war. Later on he was admitted into the Economics Department of the country's leading public university, Dhaka University, where he earned a Master's degree in economics. He has 25 years of experience doing research on many social development issues, such as poverty alleviation, empowerment of women, disability and aging, land reform, water management, and water culture.

Kazi Rozana Akhter has Master of Science (Demography), and Master of Arts (Philosophy) degrees. She has more than 15 years of experience in community-based fisheries management projects, participatory planning studies, NGO-government partnerships, disaster response, and evaluation research in water-related and sanitation projects and others. She has published a study of Bangladeshi women's responses to floods. She is a Senior Associate of Planning Alternatives for Change.

Shireen Akhter has Master of Social Science (Sociology) and Master of Science (Anthropology) degrees, and more than 15 years of experience with training, social research, program planning and program implementation in Bangladesh. Her specialties include water and sanitation, gender and development, and poverty. She has published an evaluation study of a water and sanitation program in urban slums. She is a Senior Associate of Planning Alternatives for Change.

Anwar Islam, who has a Bachelor of Science degree in Sociology, is a researcher with more than 15 years of experience in development related studies, programs, and projects, including several relating to water and sanitation. He has served as a senior researcher in our water culture study, doing ethnographic and folklore studies in his home area of Delduar, Tangail, Bangladesh. He is an Associate of Planning Alternatives for Change .